JOURNAL OF RESEARCH

OF THE

AMERICAN FEDERATION OF ASTROLOGERS

Vol. 13

Editors: James Herschel Holden, M.A., FAFA,
Research Director of the A.F.A.

and

Kris Brandt Riske, M.A., PMAFA,
Executive Director of the A.F.A.

Funded Through a Grant
from the Catharine and Ernest Grant Trust

ISBN-10: 0-86690-607-X
ISBN-13: 978-0-86690-607-4

Cover Design: Jack Cipolla

Published by:
American Federation of Astrologers, Inc.
6535 S. Rural Road
Tempe, AZ 85283

www.astrologers.com

Printed in the United States of America

TABLE OF CONTENTS

Editors' Preface v

Combustion: An experimental Study in scientific astrology . 1
Edouard Symours

Blending Meteorology and Astrometeorology 45
Kris Brandt Riske

60 Year Intervals and October Panics 57
David McMinn

Distant Planetoids 65
James Herschel Holden

Eleven Police Officers Killed while on Duty 75
James Herschel Holden

What will Happen Next? 79
James Herschel Holden

The Horoscope of the Emperor Tiberius 111
James Herschel Holden

Claudius Ptolemy 121
James Herschel Holden

The Purpose of the Trepidation of the Equinoxes 131
James Herschel Holden

Book Reviews 137
James Herschel Holden

Obituaries 141

EDITORS' PREFACE

This issue of the *Journal of Research* has been considerably delayed due to the indisposition of the previous editor and the recent reorganization of the AFA Headquarters Office. Some papers that were submitted earlier are not yet available and may be published in the next issue. However, we have a number of papers on hand for the current issue, as well as some other items of interest.

We begin with an important paper by a French astrologer on the subject of Combustion. And we continue with papers on Meteorology and Astrometeorology, and Market Cycles, followed by several other papers on a variety of subjects, a dozen book reviews, and sadly some obituaries.

James Herschel Holden, FAFA, Research Director
Kris Brandt Riske, PMAFA, Executive Director

EDOUARD SYMOURS

COMBUSTION

AN EXPERIMENTAL STUDY
IN SCIENTIFIC ASTROLOGY

Translated from the French

by

James Herschel Holden, M.A.
Fellow of the American Federation of Astrologers

CONTENTS

Translator's Preface 5
French Title Page 7
Author's Dedication 9

Combustion by Édouard Symours

Introduction 11
Combustion, An Experimental Study 21
Observations on the Statistics
I. Longevity 26
II. Tuberculosis 28
III. Violent Death 29
IV. Suicide 30
V. Criminals 31
VI. Madness 31
VII. Derangement 32
VIII. Learned Men 33
IX. Artists 34
X. Politicians and Military Men 35
XI. Celebrities 35
General Summary 36
XII. Fatal Illnesses 39
XIII. Violent Death 40
Conclusions 41
Orbs 41
Applications and Separations 42

Translator's Preface

The only extensive study of combustion that I know of was made by the French astrologer Édouard Symours and published by Éditions des Cahiers Astrologiques at Nice, France, in 1946. So far as I am aware, it has not been reprinted nor has it been translated into any other language. I have chosen to translate the full text of the results that Symours obtained and recorded on pp. 35-55 of his book. The earlier part of the book is historically interesting, and it contains on pp. 22-34 the numerical statistical tables, but the portion that I have translated sets forth the author's findings from these tables adequately.

Those who can read French may wish to obtain a photocopy of the earlier pages of the original work containing Symours's Preface (in which he discusses the importance of statistics in astrological research) and the tables of numerical data.

I shall define some terms that Symours uses in discussing his findings. For each group of horoscopes, the term "special frequency" refers to the actual percentage of cases found with the combustion under discussion. The term "general frequency" refers to *an expectancy* that Symours established by determining the percentage of cases of that combustion found in a "general group" of horoscopes that was much larger than the group under study. This is not exactly the same as the "astronomical expectancy" modified by demographic factors, but Symours believed it to be reasonably close to that and therefore satisfactory for comparison purposes. Since the total number of horoscopes he studied was 2,824, the general group chosen from them was fairly large. But it would be more accurate to say that Symours's results come from the comparison of a particular smaller study group with the much larger total group. Nevertheless, the results are an important first effort to establish the observed effects of combustions.

James Herschel Holden
October 2002

EDOUARD SYMOURS

LA COMBUSTION

ÉTUDE EXPÉRIMENTALE D'ASTROLOGIE SCIENTIFIQUE

ÉDITIONS DES CAHIERS ASTROLOGIQUES
15, Rue Rouget-de-l'Isle — NICE

This Study is Dedicated to the Memory of Paul Choisnard,
Renovator of Scientific Astrology, to Whom I am Indebted
for What is the Best of my Works.

INTRODUCTION

TRUE AND FALSE ASTROLOGY

Madame Blavatsky announced in one of her works that the twentieth century would see the renaissance of astrology.

This prediction has been realized.

Astrology is now à la mode. It is spoken of on all sides, and many devote themselves to it.

We would like to be able to congratulate you without reserve for that current which drives spirits towards the study so passionant and so useful of the celestial influences. Unfortunately, as often happens in such circumstances, this movement in favor of astrology shows some undesirable and sometimes even deplorable sides.

First of all, there is the "commercial charlatinism that permits some unscrupulous individuals to exploit brillantly a "rich vein," for human creulity is unlimited!

In our country, anyone can, without any knowledge of the matter, proclaim themselves "astrologers," or even "mages" or "fakirs," and fill their pockets at the expense of the naïve and the credulous attracted by a flashy publicity. Few people question the difference that can exist between an expert scientific astrologer and a fake.

All the charlatans that we are denouncing ridicule and dishonor the most magnificent of the sciences, "the most beautiful gift that heaven has given to the earth," an Oriental has written. It is simply scandalous that they are able to practice their profession of swindler freely without being hindered by the public authorities. The affair of the fakir "Birman," who had won millions by exploiting innumerable dupes, and who was condemned for his fraudulent manoeuvres, constitutes a good example of "this kind of industry," which has no relation with the true science of the stars!

One single country (where besides, scientific astrology takes a large extension) had the courage to take some measures that were imperative. By eliminating the swindlers and giving an "official status" to scientific astrologers who alone were authorized to practice this science after having undergone an examination testifying to a deep knowledge and in conditions that recall the medical profession among us, that country has known how to put astrology in its true place.It will be painful for you to know that that country was Hitlerian Germany.[1]

The sore of "astrological charlatanism" that we are denouncing is not the only one. Among astrologers there are also some people who are not entirely safe from reproach.

Many beginners who have only a superficial knowledge of the question, after having read some books and cast some horoscopes, are putting themselves forth to "sell horoscopes" and even, a thing more serious, to give lessons in astrology! One fears the results!

If we pass now to astrologers having a more complete *theoretical* knowledge, we are going to encounter several "reefs" that it is necessary to have the courage to mention.

French literature of the Grand Siècle had the inestimable privilege of possessing in the person of Boileau[2] a judicious critic, full of good sense and judgment. He put everyone in his own proper place, counseling and encouraging the writers of value and chastising pitilessly, with a keen vivacity, the "scribblers" without talent who were never able to rid themselves of their unavoidable hindrances.

Astrology would have a great need of having its own "Boileau"

[1]It is true that at its decline the Hitlerian regime persecuted astrologers, just as the Roman Empire once did!

[2]Translator's Note. Nicolas Boileau-Despréaux (1636-1711), famous French poet and critic.

to pass through the sieve of an enlightened judgment all those authors who write on it and to make a judicious sorting between the good and the bad.

We are too well aware, alas, of those constructers of marvelous systems who forge from all the pieces in accordance with what they call their intuition or their esoteric views some methods of interpretation that are new as well as erroneous and who make themselves out to be "masters" as viewed by the novices and the ignorant. These are, to use an evangelical term, "the blind leading the blind."

Even among those astrologers who are well-known and capable, there is a malady that is too often as wide-spread as this, consisting of constructing and defending theories that flatter their personal pride without worrying about knowing whether or not they square with impartially observed facts (which is perhaps the only thing that counts!).

And if one tried, with proof in hand, to bring them back to a respect of the realities, they will perhaps answer like that philosopher, of whom Henri Bergson tells us in his *Rire*, the pleasing answer to the objections by which it was shown to this audacious debater that the facts were contradicting his theories: "Oh, well! Experience is wrong," he says, to put an end to the discussion!

However, there exists a simple and infallible procedure to lead astrological research to the discovery of truth, which is to verify the traditional rules or to bring to light some new astral correspondences.

This procedure has been set forth in all its details by Paul Choisnard,[1] the brilliant renovator of astrology. It is the application to astral science of the statistical method based on the calculation of probabilites. Here is how Choisnard defines the "astral correspondence":

[1]Translator's Note. Paul Choisnard (1867-1930), the pioneer in the application of statistical research to astrology.

> "In the sky at birth, an astral aspect is said to correspond to such and such a human faculty (in character or in destiny) when that aspect is encountered *more frequently* among men endowed with that faculty than among other individuals."

The astrological fact under consideration is then only fundamentally a question of the deviations of frequencies to establish experimentally by means of compared statistics.

A concrete example will permit us to comprehend how it is necessary to proceed. Let us suppose that we would like to determine whether the conjunction of Venus with the degree of the ASC is favorable to longevity.

First of all, we are going to gather together a respectable amount of birth data relating to some persons dead at an advanced age (for example, between 90 and 100 years or more).

After having cast the horoscopes, we will note for each one of them the angular distance (in degrees) between the ASC degree and Venus, and we will select all the cases in which these two astrological factors are in conjunction; that is to say, we will utilize an orb of influence of the chosen aspect (5 degrees or 10 degrees both before and after, for example).

The figure found for the persons having this configuration is termed the *special frequency*.

Next, we shall take at random without any preliminary selection a sizable number of birthdates of some individuals, and we shall select the number of cases in which the same aspect exists (the conjunction between Venus and the ASC degree). The figure found will represent what is called the *general frequency*.

It would also be useful to have the control of noting the *astronomical frequency*, that is to say, the number of aspects that normally conform to the celestial movements.

That calculation is very easy.[1] Given the fact that the ASC runs through the entire zodiac in 24 hours, and that Venus, during that

[1]Translator's Note. Not if it is done correctly!

same time, moves a maximum of a degree and a quarter, if we adopt an orb of conjunction of 10 degrees in each direction, i.e. 20 degrees in all, the number of chances that one would have to fall upon a conjunction of Venus and the ASC degree would be 20/360, or around 0.055.[1]

Nearly always the astronomical frequency and the general frequency taken on a sufficiently large number of cases are very close. However, Choisnard has explained that this is not an absolute rule; and he has shown that the special frequency should not be compared with the astronomical frequency but rather with the general frequency.[2]

To pursue the example that we have chosen, let us suppose that the general frequency taken on many hundreds of cases is 5% and that the special frequency is 10%.

The ratio between the two frequencies is equal to 10/5 or 2. In other words, the number of aged persons possessing the conjunction of Venus with the ASC degree in their horoscopes is two times greater than that of ordinary individuals. There is an astral corre-

---

[1]Translator's Note. This simple procedure would only be approximately true for persons born on the equator. In moderate and northern or southern latitudes, the ASC degree moves at a variable rate through the zodiac, so depending upon the latitude and the position of Venus in the zodiac, the length of time required for Venus to pass through 10° below the horizon to 10° above the horizon would vary considerably. It was left to François Gauquelin (1928-1991) to make accurate calculations of this sort.

[2]Translator's Note. Actually, the computation of an "expected frequency" would begin with calculation of the astronomical frequency (which would vary from year to year) and then some effort would need to be made to apply a demographic correction to it (because births do not occur at an even rate through the days of the year, and the variation is both time and place dependent). It is difficult to determine an accurate "expected frequency." And such a determination also depends upon the group of persons to be studied. Symours's "general frequency" is a different approach to the problem—in effect, he is actually comparing one group of persons, say long-lived persons, against a much larger group of persons of all sorts. This is a valid procedure, but it needs to be understood for what it is.

spondence there. The greater or lesser difference between the two kinds of frequencies shows the greater or lesser importance of the law of astral influence.

As has been seen, the procedure employed is neither mysterious nor complicated, at least in its principle,[1] for the handling of statistics is often a delicate matter, and its conditions of validity are rigorous.

Be that as it may, astrology is, above all, an experimental science (which does not mean that it is not at the same time a spiritual science, very different from the official sciences on many points).

It would not be known in every case how to deal with the observation and with an impartial and rigorous control of the facts.

We admit that pure logic, intuition, esotericism, and other sources can formulate some hypotheses, but it is always necessary to have recourse to the facts to verify them (and consequently, under a more or less clear form, [to do so] in a statistic of compared frequencies).

Lacking the ability to register and measure—with physical apparatus—the astral influx, or to possess a "spiritual view," the only instruments that can help us to make astrology progress, that is to say to verify the traditional rules and to discover some new ones, are:

Either by practice by way of repeated examples, or by the statistical method. It should be said that basically statistics is always employed in both cases, but in the first, it is a question of a statistic based upon "opinion," while in the second case, one proceeds by a rigorous and impersonal control of the figures in accordance with the principle of compared frequencies.

Moreover, this principle is at the base of all the modern sciences, as it is in our judgment.

The famous Laplace[2] said once that "the study of probabilities was only common sense reduced to calculation."

---

[1]Translator's Note. Unfortunately this statement is not true. (See the preceding note.)

[2]Translator's Note. Pierre Simon, Marquis de LaPlace (1749-1827), French

This question of the probabilities in the science of observation has entered little by little into "official" science, in which it now holds an important place. That is why Mr. A. Rey wrote in the *Revue philosophique* (May 1923):

> "The introduction of the calculus of probabilities into the elementary foundations themselves of physical systematization and as a method of interpretaion of the laws of physics is a scientific event that is as important as the extension of the principle of relativity."

Here are two more interesting quotes taken from some contemporary scientific writers.

The first is extracted from the "Calculus of Probabilities for the Use of Everyone" by Messers Fréchet and Halbwachs, members of the university:

> "We have agreed to consider probability as a *purely experimental notion*, the notion of a physical constant, whose approximate measurement is obtained by the positive observation of the frequency of a fact."

The second comes from a Preface to the translation of Einstein's "Theory of Relativity" by the mathematician Emile Borel:

> "When there has been established between a system of formulas and a set of natural phenomena some relations of such a nature that the calculations based on the formulas permit us to predict the phenomena, that system of formulas constitutes a true theory, independent of any interpretation. If the experiments by which the truth of the theory has been confirmed are correct, it can be further affirmed that the theory will remain true."

From all that which has been said, it follows that scientific as-

astronomer and mathematician. He wrote a major treatise on probability, *Théorie analytique des probabilités* (Paris: Courcier, 1812) that summed up all the work of his predecessors and laid the foundations of modern probability theory. This was followed by a more elementary exposition of the subject, *Essai philosophique sur les probabilités* (Paris: Courcier, 1814). It is to one of these that Symours refers.

trology, such as that which Choisnard has defined and practiced, imposes itself on every impartial intelligence with the same standard as all the other contemporary sciences based upon observation and experimentation.

The method inaugurated by that eminent savant has already borne some magnificent fruit in France and abroad. And yet, a thing scarcely believable, there is a large number of astrologers who never miss a chance to express their contemptuous disdain for statistics, that matter of fact process that according to them leads to nothing! (Note 1).

These gentlemen have, fortunately, passed over this lower stage and on the wings of intuition and their personal experience they have taken their flight towards the heaven of absolute truth!

But what is regrettable is that when these great spirits expose the results of their "researches" and when one wants to take the trouble to review them with the fine-toothed comb of methodical experience, their theories collapse like a house of cards.

How could it be otherwise?

However, we do not deny anyone the freedom to choose his own pathway.

For our part, we have followed from the beginning of our work (1925) the one that the lamented Choisnard knew how to lay out so masterfully. We have been satisfied with it everywhere, for it permits us to extend in an unhoped for manner the field of astral correspondences that are verifiable experimentally. The results thus obtained would require a large volume to be set forth in their totality.

The modest study that we are presenting here is only a small fragment of a whole series of works approaching astrology in its most diverse aspects. It certainly shows some lacunas and exposes itself to a further critique; but we do not think that the method of investigation utilized (compared frequencies) can be contested in principle . . . and that is the essential thing.

If the results which we have achieved differ from the theories of certain of our confreres on the effects of combustion, let them not hold us rigorously to it, for we have only tried to follow experience as our sole guide, sometimes even sacrificing our own conceptions.

We can, however, legitimately invoke the old and wise saying: "Amicus Plato, sed magis amica Veritas!" [Plato is a friend, but Truth is a greater friend]

E. S.
January 1945

NOTE 1. For those of our readers who would like to have an overall view of Choisnard's work, we recommend that they read the work of the Viscount de Herbais de Thun, *Synthèse de l'oeuvre de P. Choisnard.*

We are happy to take this opportunity to address to that eminent pioneer of scientific astrtology the expression of our warmest thanks for the very devoted cooperation that he has generously accorded us.

COMBUSTION

(An Experimental Study in Scientific Astrology)

by Édouard Symours

It is said that a planet is "burnt" or "combust" when it is found close to the Sun. The orb of combustion does not seem to be very precisely indicated in the traditional works. Some of them fix it at three degrees; but in general the assigned orb is more extensive and goes from eight to ten degrees. In Gouchon's dictionary,[1] which contains some valuable observations on the traditional rules, the orb of combustion is eight and a half degrees.

A planet is said to be in *cazimi* or in the "heart" of the Sun when it is found to be less than 16 minutes from that luminary.

According to tradition (or at least what is said to be the tradition!) combustion would always be evil, while *cazimi* would always be favorable (which appears to be a bit strange and illogical, since *cazimi* is only a combustion intensified to its maximum!).

The traditional and modern authors are in general accord in saying that combustion is ominous (however, Gouchon, who has a lengthy experience in interpretation, assumes that at least for Venus and Mercury, combustion is often favorable). However, the opinions vary on the effects produced by combustion and on why combustion produces them.

---

[1]Translator's Note. Henri Gouchon (1898-1978), *Dictionnaire astrologique* (Paris, 1936-1940. 3 vols.), the standard French astrological dictionary.

[2]Translator's Note. Antarès, pseudonym of the Belgian astrologer Georges Mostade (1900-1988).

Antarès[2] says that in the case of combustion the effects manifest themselves in a manner that is excessive and pernicious, brusque, exaggerated and precipitous.

According to a study that appeared in 1932 in the *Almanach Chacornac*, combustion would deprive the planets of their personal action, which the Sun would monopolize.

Tradition, according to Gouchon, attributes to combustion an enfeeblement, a "vitiation" (unless the planet is in *cazimi*).

Janduz[1] expresses herself thus: "Combustion, a conjunction with an orb of 1 to 8 degrees, absorbs the benefic planets as well as the malefic ones. It is as if the burned planet no longer exists, at least with regard to its beneficent effects; because, for unfortunate things, combustion aggravates the presages, especially if the Sun is conjoined to a malefic."

In truth, this question of combustion is complex, leaving aside the cases of eclipses and occultations, which are of great interest in mundane astrology, but which one only occasionally encounters in a natal horoscope that is to be interpreted.

The *cazimi* is likewise quite rare, and it is necessary to examine a sizable number of horoscopes to encounter any examples of it.

Combustion is simply a conjunction to which a particular importance is attributed because of the nature of the Sun, the center and heart of our astronomical system . . . and we could apply the rules of a conjunction to it. It is true that the theory of conjunctions is very much at issue and has made much ink to flow without one being able to see the matter clearly (let us say in passing that we have personally dedicated some important experimental research to this subject and that according to the results obtained, most of the moderns who have written on this question[2] have committed grave errors due to their having neglected the facts in favor of theory).

---

[1]Translator's Note. Janduz, pseudonym of Jeanne Duzea (c.1895-c.1965), author of several astrological works, including an encyclopedia.
[2]Caslant, among others, is gravely deceived in affirming that the conjunction is always evil. Translator's Note: Col. Eugene Caslant (1865-1940), was a leading figure in the French astrological revival.

In combustion, there is a distinction to be made between the superior planets and the inferior planets (Venus and Mercury).

We note that for Venus and Mercury it is proper to take into account the superior and inferior conjunctions which are very different astronomically and whose effects could very well *a priori* not be alike.[1]

In taking into account these diverse distinctions, we are going to present the results of several statistics relative to the diverse "combustions" of the Sun.

Before this, it is indispensible to give the general frequencies (percentages) of the diverse "combustions" of the Sun with the planets.

These general frequencies, taken on several hundreds of horoscopes, which are very near in almost all cases to the astronomical frequencies (relative to the movements of the stars) calculated for several years in the ephemerides.

Choisnard has explained very well why one must compare the special frequency, not to the astronomical frequencies, but to the general frequency. Also, it is not useful for us to give here the list of astronomical frequencies, which, we say again would make a double usage with that of the general frequencies.

One will note the rarity of the inferior conjunctions of Venus. As a result, in order to be able to appraise the effects of that conjunction, several thousand horoscopes would have to be studied.[2]

---

[1]Translator's Note. The superior conjunction occurs when Mercury or Venus is on the far side of the Sun from the earth, while the inferior conjunction occurs when the planet is between the Sun and the earth and is therefore closest to the earth. The superior conjunction would thus correspond to the conjunction of the Sun with Mars and the other outer planets, while the inferior conjunction of Mercury and Venus (when the planet is retrograde) would correspond to an *opposition* of the Sun and Mars or one of the other outer planets. So, as Symours says, the effects of the two types of conjunctions of Mercury and Venus might well be quite different.

[2]Translator's Note: Averages taken from the A.F.A. *Relative Frequency Tables* (Tempe, Az.: A.F.A., Inc., 1978) over recurrence periods of Mer-

	Orb of 5°	*Orb of 10°*
Sun and Neptune	2.8	5.6
Sun and Uranus	3	6
Sun and Saturn	3.6	7.3
Sun and Jupiter	3.7	7.4
Sun and Mars	4.4	8.8
Sun and Venus inferior conjunction	1.1	2.2
Sun and Venus superior conjunction	7.2	14.4
Sun and Mercury inferior conjunction	5	10
Sun and Mercury superior conjunction	10.2	20.4

Now we are going to present the results obtained from our various statistics.

We have thought that in order to arrive at a clear presentation of the effects of combustion, it was indispensible to present these statistics on several [different] "plans."

In fact, nothing proves à priori that, for even the same planet, combustion must produce identical results, when it is a question of physical health, mental faculties, or destiny.

Experience, as one will see, justifies this kind of thinking: it is thus that the conjunction Sun-Jupiter, moderate or even dangerous for health (when it is within narrow orbs), reveals itself to be fortunate for [mental] equilibrium and the strength of the psychic faculties and good for social success (wealth, power, honors, etc.).

We have for some years made a statistic of the conjunctions of the Sun and Jupiter found from a list of birthdates of some French senators. We have been able to ascertain that the special frequency, relative to that conjunction surpassed the general frequency by more than 40%.

However, we have observed on the other hand in the course of

---

cury and Venus in the early 1900's yield the following figures: Mercury 5° orb inferior conjunction 4.5 superior conjunction 10.0; 10° orb inf. conjunction 9.4, sup. conjunction 20.4; Venus 5° orb inf. conjunction 1.1, sup. conjunction 6.7; 10° orb inf. conjunction 2.2, sup. conjunction 13.3.

our research on longevity that this same conjunction was not at all frequent, especially with a narrow orb. These facts have led us to think that the same combustion could act very differently according to the domain in question.

This is why we are presenting our statistics on three [different] plans:

A) I.—Physical health, vitality.

B) II.—Mental faculties.

C) III.—Destiny.

Evidently, "all is in all," and it is hardly easy to isolate these different plans. For example, those persons who have a brilliant social success could not only owe it to what is called "luck," but also to good health and to intellectual qualities.

Nevertheless, there is a strong interest in proceeding as we have done, for certain results are quite remarkable.

We have also thought that it would be interesting to take into consideration different orbs, for, there again, one can estimate à priori that the extent of an orb can entail some very different results according to the case.

Here again, the facts have confirmed that supposition.

It is thus that the close conjunction of Mars and the Sun is as dangerous as possible for life, and consequently very rare in the case of longevity. On the contrary, the sufficiently separated conjunction does not have the same ominous character, and it reveals itself to be favorable enough.

For each conjunction, then, we shall give three orbs: the first, going from 0 to 5 degrees (in each sense); the second, from 5 to 10 degrees; and the third, from 0 to 10 degrees.

Finally, we have also taken note of the application and separation of the various conjunctions; and there too, the results have brought forth some lessons worthy of consideration.

To cite only a single example, which is very curious, the close combustion between the Sun and Uranus reveals itself to be frequent at times among centenarians . . . and [also] among those persons dead from a violent death; but, in the first case, only the separation is favorable, and in the second, it is the application that is menacing. . . .

Our researches have been based upon a total of 2,824 cases, which authorizes sufficiently solid conclusions.

[Following this, Symours gives tables of his statistical results on pages 22-34. I have omitted them here. Those who would like to see the tables should acquire a copy of the French original, which is unfortunately out of print, but a photocopy could be obtained. Symours's conclusions from the statistics that he amassed are contained in what follows.]

OBSERVATIONS ON THE FIRST STATISTIC

(200 cases of longevity)

1. The conjunction of the Sun and Neptune does not appear to be very significant. In fact, with a narrow orb the general frequency and the special frequency are essentially equal.

2. The conjunction of the Sun and Uranus (already signaled as an index of longevity by the astrological statistician Krafft[1]) seems in fact to play a favorable role but only when it is in separation.

In fact, when the general frequencies are respectively 3.3 and 6.6 corresponding to the orbs 0 to 5° or 5 to 10°, one obtains for separation the figures 8.7 and 15, that is to say more than double, while the applications give the figures 2.2 and 4.

The reason for this fact is no doubt due to the very violent and dynamic nature of Uranus, whose action on the Sun (and consequently on vitality and health) can be good only on the condition of being moderated (which takes place in the separation but not in the application).

3. The conjunction of the Sun and Saturn, signaled constantly by the astrology manuals as being very evil, does not seem to merit

[1]Translator's Note. Karl Ernst Krafft (1900-1945), Swiss astrologer, a pioneer in applying statistical analysis to astrology. He was the successor to Paul Choisnard in astrological research.

its reputation. In fact, with an orb of 0 to 5°, its special frequency is 6 (against 7.2 general frequency). Better than that, at a more extended orb (5 to 10°), the special frequency yields 11, of which 9 is for the application, against 7.2 for the general frequency.

It should [also] be noted that according to our statistical research, the squares between the Sun and Saturn are, on the contrary, dangerous for life.

The relatively benign nature of this conjunction of the Sun and Saturn is doubtless explained by the fact that the conjunction is not a violent aspect and constitutes a mixture. Now, with the Sun and Saturn possessing opposite natures (the Sun being masculine, positive, hot, etc... and Saturn feminine, negative, cold, etc.), their influences are balanced in the combustion, but here the contrary of the preceding case is produced. The application is better than the separation. The thing explains itself easily from the nature of Saturn, which is so opposed to that of Uranus.[1]

4. The conjunction of the Sun and Jupiter. Contrary to what is habitually taught, this conjunction, it seems, is not very brilliant for longevity (especially with a narrow orb and with application). This fact must always be explained by the same rules of polarity.

Both the Sun and Jupiter having the same nature (positive, masculine, hot, expansive, etc.), their conjunction, especially when very close, produces an "excess" of heat which can harm the

[1]Saturn harms by insolvency or by deficiency: yet the application, which fortifies it, can without doubt ameliorate it in some measure. Another curious example of the action of polarities has been furnished to us by a statistic relative to the aspects between Venus and the Moon among centenarians (especially those of the feminine sex). In fact, we have observed a considerably elevated frequency of the square of these two stars, especially of the square at 90° with application. The explanation of this apparent anomaly resides without doubt in the fact that Venus and the Moon, both of them being feminine, passive, and negative, could have needed that their influence (which is sometimes hypoactive) be reinforced to be "excited" by an aspect that is violent and dissonant, but also "dynamic" like the square. On the physiological plane, it is probable that that translates itself through the reinforcement of the action of certain glands governing the internal secretions (the Endocrines).

health. In fact, we have often observed in medical astrology that the dissonances between Jupiter and the Sun threaten [health] by congestion, apoplexy, etc.

5. The combustion of the Sun and Mars seems very dangerous, especially at close distances (the special frequency is three times inferior to the general frequency). Ptolemy of Pelusium has already signaled the "dangers of Mars, who is contrary to the Sun" (*Tetrabiblos* i. 16).[1]

The preceding explanation is still acting here but with more force because Mars is particularly "blazing and excessive."

One notices by contrast that the more elongated conjunction does not seem unfavorable, and we have noted especially between 10 and 20 degrees of distance a number superior to the general frequency among our centenarians.

6. The conjunction of the luminaries (i.e. the new Moon) seems ominous, especially at small distances (only one case between 0 and 3 degrees orb in a group of 200!).

7. The superior conjunction of Venus with the Sun appears to be distinctly favorable, but [only] on the condition that the orb is small (0 to 5°).

8. The superior conjunction of Mercury seems to be slightly favorable. On the contrary, the inferior conjunction seems to act in the opposite sense. (But the number of cases considered hardly permits one to conclude [with certainty].)

OBSERVATIONS ON THE SECOND STATISTIC

(100 cases of pulmonary tuberculosis)

In order to show more clearly the favorable or unfavorable influence of the several combustions, it would have been necessary

[1]Translator's Note. *Tetrabiblos* 1. 16 ("On Disjunct [Signs]" in Robbin's ed. and "On Seeing and Equal-powered [Signs]" in the Teubner eds.) is about the signs and says nothing about Mars and the Sun. Evidently Symours wrote down the wrong place in the *Tetrabiblos*. I have not found the place he cites.

to use [only] the horoscopes of those persons who succumbed prematurely. But our list contains a goodly number of tubercular patients who have resisted the malady, which diminishes the "relief" of the present Statistic.

This Statistic constitutes in some way a "contra proof" of the preceding Statistic.

On the whole, the results confirm those obtained previously. The combustions of Neptune and Uranus seem to be devoid of any noticeable action. Those of Saturn and Jupiter offer some general frequencies very close to the special frequencies. It is interesting to note that the Sun-Jupiter conjunction does not have any "protective" effect.

The combustion of Mars is clearly dangerous, but only with a narrow orb and especially with application. Outside of the 5° orb, this conjunction no longer seems to be harmful, but on the contrary good.

The conjunction of the lights is doubtful.

The combustion of Venus is rare with a narrow orb (it is then a favorable indication for health).

Both the superior and inferior conjunctions of Mercury seem to be without any great signification.

OBSERVATIONS ON THE THIRD STATISTIC

(400 cases of violent death)

1. The conjunction of the Sun and Neptune seems to be dangerous (especially with narrow orbs and with application). One will note in passing that the dissonances of Neptune with regard to the vital significators correspond rather often to demises by submersion (radical aspect, transits, and directions).

2. The conjunction of Uranus and the Sun is more dangerous still, especially [when] close and in application.

3. The combustion of Saturn appears little active with narrow orbs, and, with wider orbs, it would be rather protective against a

violent death (which is logical, given the nature of Saturn).

4. The combustion of Jupiter seems to be rather "protective," (even with narrow orbs), but with the condition that this is in separation.

5. The combustion of Mars is definitely dangerous (especially with narrow orbs and in application).

6. The same observation is valid for the conjunction of the lights.

7. The general frequency of the Venus-Sun conjunction is close to the special frequency, but it should be observed that application is twice as frequent as separation (but among centenarians, on the contrary, separation was more frequent than application).

8. Conjunction of Mercury-Sun. The special frequency of the superior conjunction slightly surpasses the general frequency. It should be noted again that application is twice as frequent as separation.

For the inferior conjunction, the special frequency is a little bit inferior to the general frequency, but application and separation are equal [in number].

OBSERVATIONS ON THE FOURTH STATISTIC

(100 cases of suicide)

1. The conjunctions of Neptune and the Sun with a narrow orb are dangerous (special frequency double that of the general frequency).

2. The conjunctions of Uranus and the Sun with a narrow orb are equally dangerous (the same frequencies as previously).

3. The Sun-Saturn conjunction that we have found rather benign among centenarians and in the cases of violent death seem this time to be very characteristic (the special frequency is double that of the general frequency). It is to be presumed that this aspect can cause a vital depression and melancholy, often provocative of self-destruction.

4. The Jupiter-Sun conjunction by contrast (and it is logical) is very rare among suicides—a single case out of 100 horoscopes. It is evident that this aspect acts mentally and physically in a sense that is entirely opposed to that of the preceding aspect.

5. The close combustion of Mars is dangerous as was expected.

6. There is nothing special about the Venus-Sun conjunction (the special frequency is close to the general frequency).

7. The superior conjunction of Mercury shows a special frequency that is a little less than the general, and in particular there are three times more separations than applications.

8. There is nothing prominent for the conjunction of the lights.

OBSERVATIONS ON THE FIFTH STATISTIC

(100 horoscopes of criminals, most of them executed)

1. For Neptune and Uranus: combustions with a narrow orb are alarming (special frequency double the general frequency).

2. The combustion of Saturn seems to be rather unimportant.

3. The combustion of Jupiter, whether close or not, is rare. This aspect favors healthy morals and destiny.

4. There is nothing very clear for the combustion of Mars.

5. There is nothing particular for the new Moon.

6. The conjunctions of Venus and Mercury are remarkable for their rarity (especially with a narrow orb).

The special frequency is 6 (vs. 15.2) for Mercury, and 3 vs. 8.3 for Venus; it is doubtful then that these aspects favor mental equilibrium and "humanizing" the intellect.

OBSERVATIONS ON THE SIXTH STATISTIC

(250 *cases of various kinds of madness)*

1. There is no clear action from the combustions of Neptune and Uranus.

2. The conjunction of Saturn does not seem to be very influen-

tial: it is somewhat "benign." (Saturn gives some "weight" to the intellect.)

3. The same observation applies to the conjunction of Jupiter. However, with a wide orb it seems to be good.

4. The combustion of Mars is evil.

5. That of the lights is doubtful.

6. The close combustion of Venus is rare (13 vs. 18).

7. The combustion of Mercury presents itself differently, according to whether it is the case of a superior conjunction (mean frequency) or an inferior conjunction (four times below the general frequency).

OBSERVATIONS ON THE SEVENTH STATISTIC

(632 *cases of derangement or dementia praecox)*

1. For Neptune and Uranus, nothing very clear. However, the special frequency would tend to be slightly greater than the general frequency.

2. For Saturn the special frequency is a little lower than the general frequency (38 vs. 46). Saturn then protects the mental equilibrium somewhat.

3. The combustion of Jupiter offers a very remarkable result. The special frequency is 25 instead of 46.5, nearly half. One can then judge that this conjunction protects the mentality against the "dissociation" engendered by dementia praecox.[1]

4. The Sun and Mars: the special frequency is very slightly greater than normal.

5. The conjunction of Venus is remarkable. In fact, one finds only 72 cases vs. 104.5.

6. The same is true of Mercury: the special frequency is 115 vs. 191.5. It is then clear that the Mercury-Sun conjunction protects

[1]Jupiter's relation to the "hierarchy of functions" and their coordination would seem here to be very suggestive, if one likens it to the dissociation engendered by dementia praecox.

mental equilibrium.

7. The Sun and the Moon: there is nothing in particular to note.

This statistic demonstrates to us as characteristic elements the rarity of the combustions of Jupiter, Venus, and Mercury. If the results from Statistic 7 give more clear deviations of frequencies, this fact is due to the particularly serious and incurable character of dementia praecox, whereas in the preceding statistic there are some mental troubles of all kinds and of very variable severity.

Be that as it may, it clearly results from these researches that, far from being inauspicious, the combustions of Mercury and Venus are on the contrary very fortunate and favorable to the psychic equilibrium.

OBSERVATIONS ON THE EIGHTH STATISTIC

(210 *horoscopes of learned men and philosophers)*

1. The combustions of Neptune, Uranus, and Saturn give special frequencies close to the general frequencies.

2. The combustion of Jupiter would seem to be favorable.

3. Mars combust seems to be slightly unfavorable.

4. The conjunction of Venus and the Sun (with the condition of narrow orbs) is very good, especially the superior conjunction, and in particular the separation.

5. Both the superior and inferior conjunctions of Mercury are clearly good. (The former more than the latter.)

An interesting note: the tradition attaches a certain importance to the position of Mercury before or after the Sun (i.e. when it is oriental or occidental). The oriental position was considered to be particularly brilliant for the mental faculties and spirit.

The present Statistic seems to support a confirmation of the tradition, since Mercury oriental (before the Sun) gives for each type of conjunction (inferior and superior) a special frequency that is more than double the general frequency.

It is interesting to recall, by way of comparison, our Statistic 4 (violent deaths), which also throws into relief the oriental position of Mercury (two times more frequent than the occidental).

But on the contrary, in Statistic 5 (suicides), Mercury is more often occidental than oriental. And there is the same result for cases of tuberculosis (Statistic 2).

With reservations pending further research, it seems that one can consider that *Mercury oriental* is more powerful than Mercury occidental.

OBSERVATIONS ON THE NINTH STATISTIC

(300 *horoscopes of artists)*
100 poets, 100 painters, 100 musicians

1. The combustion of Neptune is quite favorable, which is logical because Neptune is the planet of imagination and intuition.

2. The combustion of Uranus seems to be still more favorable. One knows that this planet strongly influences the gift for music and that it acts upon the originality of mind.

3. There is no noticeable influence from the combustions of Saturn, Jupiter, or Mars.

4. The conjunction of the lights seems to be favorable; no doubt it excites the imaginative faculties, does it not?

5. The conjunction of Venus (the superior conjunction) is very good, but it only acts with narrow orbs. The number of cases is not big enough to be able to judge the strength of the inferior conjunction, which seems to be less brilliant?

It should be noted for the superior conjunction that Venus in application to the Sun (i.e. oriental) gives a special frequency double the general frequency.

6. The superior conjunction of Mercury and the Sun is excellent (special frequency 44 vs. 30.6 for the general frequency).

But again, this conjunction only has strength with narrow orbs.

It seems that the inferior conjunction is less brilliant, but it seems to act with wider orbs.

OBSERVATIONS ON THE TENTH STATISTIC

(150 *horoscopes of politicians and military leaders*)

1. The close combustion of Neptune seems to be favorable.

2. That of Uranus and that of Saturn would seem to be without any obvious influence.

3. The combustion of Jupiter is rather good.

4. The combustion of Mars, mediocre with narrow orbs, seems fairly good with a certain extension (above 5-6 degrees).

5. The conjunction of the lights is bad. It seems to injure the faculties of a man of action.

6. The conjunction of Venus seems to be only slightly favorable.

7. The superior conjunction of Mercury is excellent (24 vs. 15.3), with the condition that the orb is narrow.

The inferior conjunction seems to be hardly active.

OBSERVATIONS ON THE ELEVENTH STATISTIC

(300 *celebrities of various sorts)*

1. The close conjunctions of Neptune and Uranus are rather favorable.

2. The conjunction of Saturn is rather unfavorable, but not so clearly as one would like to think.

3. The combustion of Jupiter is rather good (but not so good as one would have thought it to be à priori!).

4. The conjunction of the lights is clearly rather hostile.

5. The superior conjunction of Venus is good (3 against 21.6), but only with narrow orbs

The inferior conjunction seems to be less effective.

6. The superior conjunction of Mercury is very fortunate (47 against 30.6), but the inferior conjunction does not seem to act in a clearly favorable manner.

General Summary of the Several Combustions According to these Statistics.

It is both interesting and useful to summarize in a clear and precise manner the principal influences of the planets in combustion on various features of life:

A. Physical health and vitality. Accidents.

B. Mental faculties.

C. Destiny.

Neptune.

A. The influence on physical health is hardly noticeable; but there is rather often the threat of accidents (violent death).

B. Somewhat dangerous from the point of view of mental equilibrium (see Statistics 4, 5, 6, 7: Suicides, Criminals, Madness). However, this combustion frequently favors the psychic faculties (Statistics 8, 9, 10: Learned Men, Philosophers, Artists, Men of Action).

C. Rather favorable for success, in general, although there are some dangers and fluctuations.

Uranus.

A. A sufficiently clear influence on life and health.

It favors longevity (but only when separating). The application is dangerous (violent death).

B. The observations are analogous to those concerning Neptune. However, the combustion of Uranus especially favors the artistic faculties.

C. The influence is comparable to that of Neptune.

Saturn.

A. The influence is little marked: slightly unfavorable. It hardly inclines to violent death (with the exception of suicide).

B. It does not seem to stand in the way of intellectual faculties. Rather, it would protect the intellect against the lack of equilibrium; however, it inclines to suicide, doubtless because of vital depression and melancholy.

C. The influence is rather marked, but slightly unfavorable.

Jupiter.

A. In close conjunction it is not favorable for physical health (various kinds of dangers), but the wider conjunction is better; it protects against violent death.

B. A very fortunate action on mental equilibrium (it is rare among the deranged, suicides, and criminals). It is rather good for the flights of various kinds of intellectual faculties.

C. Rather favorable for success on the social plane. (But the trine, sextile, and quincunx aspects seem in general to be more fortunate than the conjunction.)

Mars.

A. Very dangerous for life with narrow orbs (especially below 3°). It stands in the way of longevity and often tends to produce a violent death. However, the widest conjunction acts in an opposite manner and seems to be rather favorable in most cases.

B. Its action on mental equilibrium is rather injurious, less so than on the mental plane; but its influence on the intellectual faculties does not seem to be clearly unfavorable.

C. Slightly unfavorable. (It produces blows, struggles, reverses, etc.)

The Moon.

A. Rather bad for longevity; danger of violent death.

B. Rather well marked action, rather unfavorable for equilibrium and the richness of mental faculties (with the exception of the imaginative and sensitive qualities—see the Statistic relating to artists.).

C. Generally unfavorable for social success.

Venus.

A fortunate influence from every point of view, but it only acts with narrow orbs.

A. It favors longevity; it protects against illnesses and violent death (especially when separating or in its inferior conjunction).

B. Excellent for mental equilibrium (a protection against suicide, criminality, and madness; in the latter case, it is especially the inferior conjunction that is excellent.).

A very fortunate influence on the intellectual faculties (science, philosophy, and the arts), but less influence on the active faculties.

C. A propitious influence for social success and its results.

Mercury.

A fortunate influence in all domains, but it only acts with short orbs.

A. Rather good for physical health, but it does not protect against accidents and violent death (especially when Mercury is applying to a superior conjunction).

B. It protects the mental equilibrium (rare among suicides, criminals, and especially deranged persons. From this point of view, the inferior conjunction seems to be especially fortunate). It is excellent for all intellectual faculties (sciences, arts, etc.) and for action.

The application is the most favorable; but being the most powerful, it is more dangerous (violent death, madness).

C. A good influence for social destiny (especially the superior conjunction).

The combustions of Pluto and also those of the Part of Fortune and the lunar nodes are lacking in this summary table. It is not at all impossible that one might find some results for them worthy of interest. Nevertheless, this study already permits one to see a little more clearly a subject that is badly understood.

It provides some indications of the influence of various combinations. However, one point remains to be examined: What modifications could be made to these combinations from their position in the several houses and in the several signs of the zodiac?

One is in no way forced à priori to suppose that these elements are without influence.

We wanted to sound out this domain, and we present here the results of Statistics XII and XIII. We have intentionally chosen the combustions of Mars and Saturn because their natures are entirely opposed. Here are the results that we obtained:

STATISTIC XII.

Combustions of Saturn (40 cases of combustion: various illnesses that caused death)

Signs.

One notes Capricorn especially (8 cases), which is properly the sign in which Saturn has its maximum influence (according to tradition).

As for the Quadruplicities: the cardinal and fixed signs clearly win out over the common signs, which are weaker.

As for the Triplicities, one can state—which is logical—that combustion is especially inauspicious in the earth and water signs (in harmony with the nature of Saturn). The fire signs give only two cases of combustion. (Fire signs: 2; Air signs 10; Earth signs 16; Water signs 11).

Houses.

One especially notices house VII, opposed to house I, and also house V.

The general frequency of planets placed in cardinal or succedent or cadent houses is very near the special frequencies. One will note that the special frequency of the combustions that are above the horizon goes a little beyond the general frequency. But chiefly the angular combustions (within 10° on either side of the cusp of the cardinal house) are two times smaller in number than the general frequency.

STATISTIC XIII.

Combustions of Mars (42 cases of violent death)

One will note the elevated frequencies in House I and House IX.

There is little difference between the special and general frequencies for the different classes of houses.

The combustions above the horizon give a special frequency that is higher than the general frequency (26 vs. 21). The same distribution as before.

One will remark that conversely to that which obtained for Saturn, it is noted that angular positions are two times more numerous than their general frequency.

That difference could perhaps be explained in the following way: That which renders Mars dangerous is not "hypodynamism," as Saturn is notable for "hypodynamism," and for "insolvency" (if one can put it that way).

Mars is then more uneasy to the same degree that it is more "intense" in the horoscope, that is to say [when it is] more close to the angles; for Saturn, it is the other way around.

Positions in the Signs.

Leo and Libra come first.

One must note the elevated frequency of the fire signs and the low frequency of the water signs.

The positive signs (fire and air) give 26 against 16 for the negative signs.

This is the reverse of the preceding Statistic, and that is logical, for Mars is much stronger when it occupies the "positive" or masculine signs, while Saturn is much more "Saturnian," that is to say "hypoactive" when it occupies the negative or feminine signs (earth and water).

In the same way then, the "sounding" that we have given concerning the modifications given to the combustions from their positions in the various houses and in the various signs brings forth some interesting observations. They must be completed by the examination of the combustions of the other planets.

It will also be reasonable to see in what measure and in what manner the important aspects of the planets that are striking a combustion are capable of modifying its influence.

CONCLUSIONS.

The present study permits us to bring some clarity into an obscure domain, while showing us the action of the combustions of the several planets.

In the way that we have been able to state it, the question is rather complex. No general formula exists that can encompass all the combinations. One must then take into account:

I. The nature of the planets that are combust (planets that are positive, negative, masculine, feminine, hot, cold, benefic, malefic, etc.)

II. The orb.

III. Application and separation.

IV. For Venus and Mercury, the distinction between the superior and the inferior conjunction.

V. The different planes on which the action of the combustion manifests itself.

VI. The Sign and the House in which the combustion takes place, and the aspects that strike that combustion.

Laying aside the question about the particular effects of each combustion that has been developed in the present work, we shall summarize here the most important observations made on the subject of ORBS, APPLICATIONS, and SEPARATIONS. In fact, our work has permitted us to gather some interesting results relating to these problems.

Orbs.

Choisnard, as is known, used a general orb of 10° in his statistical work. This orb is sufficient in certain cases to obtain frequency differences that prove the reality of astral influences. But it is certain (the experience is there to prove it) that the use of narrower orbs is demanded if one wants to arrive at precise results.

Moreover, in certain cases, the benefic or malefic nature of a solar conjunction seems to depend upon the width of the orb: thus, the combustion of Mars down to 3 degrees is terribly inauspicious for life (0 cases among 200 centenarians!).

Down to 5 or 6 degrees this combination remains very evil, but after that it gets better, and from 6 to 20 degrees it seems to be rather clearly favorable for longevity.

There is an analogous observation for the combustion of Jupiter.

In some other cases (the great majority), it seems that the orb of effective action of a combustion does not exceed 5 or 6 degrees (One can then deduce that for the conjunctions of the other planets this value of orb would be still narrower, for the Sun is the most "powerful.")

Let us consider the numerous combustions of Venus and Mercury cited in this work: One will ascertain, in nearly all cases, that the special and general frequencies present a strong deviation between 0 and 5 degrees orb, but that from 5 to 10 degrees of orb that deviation tends to go to zero. (See especially the Statistics I, VI, VII, VIII, IX, X, etc.) The same is true of the "major planets." In general, it does not seem that the orb can exceed 6 to 8 degrees and be noticeable.

Applications and Separations

It seems that this distinction ought to be given serious consideration: One knows that according to both tradition and experience, separation is less strong than application.[1]

[1]This consideration, extended to the various aspects that the planets can form in the zodiac could perhaps furnish an explanation of the difference that exists between the effects of equal angular distances (for example, squares and trines) according to which the swifter planet approaches or elongates itself from the other. The first square or the first trine (90° and 120°) often has a very different influence from that of the second square

It results from this that in certain cases some important modifications can result from [taking note of] application or separation.

A remarkable example is furnished by Statistic I and Statistic III. In the first case it is apparent that the combustion of Uranus is good for longevity, but on the condition that there is a separation; and in the second case one observes that application is frequent in [cases of] violent death. Our Statistic of violent death is also very revealing for the effects of application. It may be judged from this:

For all the planets (with the exception of Jupiter) the number of applications is greater than the number of separations.

Applications (including Jupiter): 122.

Separations: 81.

Without Jupiter (2 applications and 9 separations), there would be:

Applications: 120.

Separations: 72.

The difference is too great and too general to be an effect due to chance: The probable deviation due to chance (Gauss's Law) is very much smaller than what one finds here.

Statistic VI (Madness) gives some comparable results (with the exceptions of Venus and the Moon). Without the Moon and Venus, we have:

Applications: 48.

Separations: 28.

Here again, the figures demonstrate quite well that it could not result from a simple "coincidence," and that we are face to face with an astrological law.

The End.

---

or the second trine (270° or 240°). Translator's Note. The classical Greek astrologers taught that the *dexter* aspects (what Symours calls the "second" aspects) were more powerful than the *sinister* ones (which he calls the "first" aspects).

BLENDNG METEROLOGY AND ASTROMETEOROLOGY

by Kris Brandt Riske, M.A.
Executive Director of the A.F.A.

Abstract: A discussion of the relations between Meteorology and Astrometeorology with an example—the recent hurricane Gustav.

Earth is a work in progress. And although the physical changes to the planet are less dramatic today than during other eras in its 4.5 billion year history, change is still the norm, not the exception. Periodic major earthquakes change the landscape, glaciers advance and retreat, and the polar ice caps go through periods of varying temperatures, of freezing and melting.

But it is weather, more than any other factor, that on a continual basis shapes and re-shapes Earth's terrain. Flooding, drought, and hurricanes impact regions, and extreme heat and severe thunderstorms spark forest fires, which ultimately result in topographical changes. On a small scale, weather can impact a community, when, for example, heavy rains trigger mudslides.

Yet even with 21st century technology it is still impossible to accurately predict the weather more than about three days in advance or to accurately predict a hurricane's path more than 24-48 hours in advance–at least by relying solely on meteorology, which has its scientific roots in the 17th century, when the barometer and thermometer were invented.

Astrometeorology, the astrology of weather forecasting, may have been used as early as 2,500 years ago when people correlated planetary movement to events on Earth. Ptolemy wrote about astrometeorology, but it wasn't until the 17th century that specific rules were recorded—by Dr. J. Goad in his *Astro-Meteorologica* and by J-B Morin in *Astrologia Gallica 25*.

So, although meteorology and astrometeorology developed somewhat simultaneously about 350 years ago, meteorology became the accepted forecasting method because of its scientific basis in math, physics, and thermodynamics. Astrometeorology, however, continues to rely on planetary alignments, for which there is still no scientific explanation.

Astrometeorology has a decided advantage: an accurate weather forecast can be produced for any date, time, or place anywhere on Earth, as well as seasonal trends and weekly hot spots. Meteorology also has its advantages: storm intensity can be determined through Doppler radar and satellite images, and measurements in the middle and upper atmosphere provide data about steering winds and the jet stream, for example.

Neither system is perfect. But by using both, by blending meteorology and astrometeorology, a higher degree of accuracy can be achieved in weather forecasting. At this point, however, there are few astrometeorologists who understand meteorology and undoubtedly fewer meteorologists who recognize the value of astrometeorology.

This article explores and discusses methods that can yield potentially significant results using both meteorology and astrometeorology in weather forecasting.

Meteorology operates on two levels: synoptic and mesoscale. Synoptic meteorology is the big picture—troughs (low-pressure systems), ridges (high-pressure systems), and mid- and upper-level winds. Included in the mesoscale are surface fronts, severe thunderstorms, tornadoes, and blizzards, as well as showers and insignificant thunderstorms and snow. When major synoptic factors are in place, they're often accompanied by mesoscale factors; both are required for major weather events such as a widespread tornado outbreak or a crippling ice storm.

Much of the weather is determined by what occurs in the middle and upper atmosphere, from about 18,000 to 50,000 feet above sea level. The 18,000 foot, or 500-mb level (a standard meteorological pressure level), is where meteorologists look for the troughs that signal weather. Troughs, which can be very deep (severe weather) or shallow (significant but not severe weather), enable meteorologists to identify areas where weather is likely to occur.

Most weather moves from west to east (in the northern hemisphere), traveling in waves, from trough to ridge to trough, and so on. This is why, for example, a storm in the Pacific Northwest moves across the country, bringing snow across the Rockies and Plains. It is simply a trough in motion, advancing eastward. And because a trough is preceded (and followed) by a ridge, weather can be fair one day, stormy the next few, and fair the next. From a scientific viewpoint, this is a simplified description of the atmosphere in perpetual motion in its ongoing attempt to equalize air pressure.

Thunderstorms are primarily surface-based, meaning that heat and moisture near Earth's surface are necessary for convection (a hot, humid day). This is also why the most severe thunderstorms occur in the late afternoon/early evening, after heat and moisture reach their maximum. When a trough is also present, severe thunderstorms, including the possibility of tornadoes, are possible. Although a shallow trough can fuel high winds and severe thunderstorms, tornadoes are unlikely to develop. (Conditions must be near-perfect for supercell thunderstorms and tornadic activity.)

When heat and moisture are present at the surface, along with a cold front, low-level convergence and upper-level divergence occur. Essentially, air is drawn in near the surface, rises within the cloud structure in increasingly powerful updrafts, and then cools, producing downdrafts. This continuous cycle of air movement results in a cumulonimbus (thunderstorm) cloud that continues to increase in size as it feeds itself.

What a trough provides is spin. As air reaches the deepest part of the trough and exits it, the air acquires maximum cyclonic (counterclockwise) spin. And spin encourages tornadoes to develop. A shallow trough produces minimum spin, while a deep trough produces maximum spin. The more spin, the greater the possibility for tornadoes, which develop on the western or southwestern end (rear flank) of a supercell in what is called a mesocyclone. (The mesocyclone often appears as a "hook echo" seen on Doppler radar.)

Although the above is an oversimplification of the meteorology of thunderstorms and tornadoes, it serves as a basis for the astrometeorology, where planets and planetary positions mirror the meteorology.

Astrometeorology involves the use of two events and their representative charts: the solar ingress and the lunar phase. The solar ingress is similar to the meteorological synoptic scale, and the mesoscale equates to both the lunar phase and the transits at the time of a meteorological event.

The solar ingress chart shows the planetary positions and alignments in effect at the beginning of the season and therefore reflects that season's weather trends. For example, cloudy, cool weather prevails throughout the season when Saturn aspects the meridian at a particular longitude. The same is true of locations where Saturn aspects the Ascendant-Descendant axis. Conjunctions and squares have the strongest effect. A meridional aspect has traditionally been viewed as the most prevalent influence, but my research shows that the Ascendant-Descendant axis can be just as significant. The meridian is north-south, following longitudinal lines from the North to South Poles. Ascendant-Descendant lines are curved, usually arcing from southwest to northeast when plotted on a map. Using this method the astrometeorologist does the same for each lunar phase, correlating the planets' places to a specific longitude and latitude, as well as to the planets' places at the same location at the solar ingress.

The solar ingress, like the synoptic scale, is big-picture, indicating the seasonal trends. Planetary positions at the ingress, especially the slow-moving outer planets, reflect areas where high pressure, low pressure, and significant storms will occur more frequently during the season. The short-term lunar phase (seven days) is closer to the mesoscale by definition. (Mesoscale weather, such as a thunderstorm, is short-term.) Using the lunar phase chart, the astrometeorologist can identify areas where storms (low pressure) and heat or cold (high pressure) will occur, which may or may not be the same as the prevailing seasonal trend.

As in meteorology, the most dramatic weather occurs when the lunar phase chart (mesoscale factors) trigger the solar ingress chart (synoptic scale factors). If, for example, a lunar phase Neptune were square a longitude where the solar ingress Neptune was conjunct the IC, the astrometeorologist would forecast a high probability of heavy rain and flooding. Similarly, if the meteorologist were to see a deep trough developing and moving into an area, along with a strong cold front, high temperatures, and an influx of very moist air, the forecast would include severe thunderstorms. The meteorologist would also use thermodynamic tools such as measurements of upward motion (rising air), convection potential, and water vapor, which are only three of more than the two dozen tools used to forecast weather.

Squall Line.

Summer 2008 saw an increased level of severe thunderstorms and tornadoes in eastern Nebraska, northeastern Kansas, northwestern Missouri, Iowa, Wisconsin and northern Illinois. Troughs continued to move through the area, and prevailing wind patterns throughout much of the summer encouraged a strong southerly flow of warm, moist air from the Gulf of Mexico.

Uranus is associated with high pressure, which can bring either extremely cold or extremely hot temperatures, depending upon the season. When in aspect with Mercury or Mars, Uranus indicates severe thunderstorms and tornadoes. From a meteorology perspective, a cold pool of air (Uranus) signals more intense convection, and Uranus seems also to be associated with the mid- and upper-level conditions, based upon my research.

At the 2008 summer ingress, Uranus was conjunct the IC at approximately 97W longitude, or just west of where the most severe

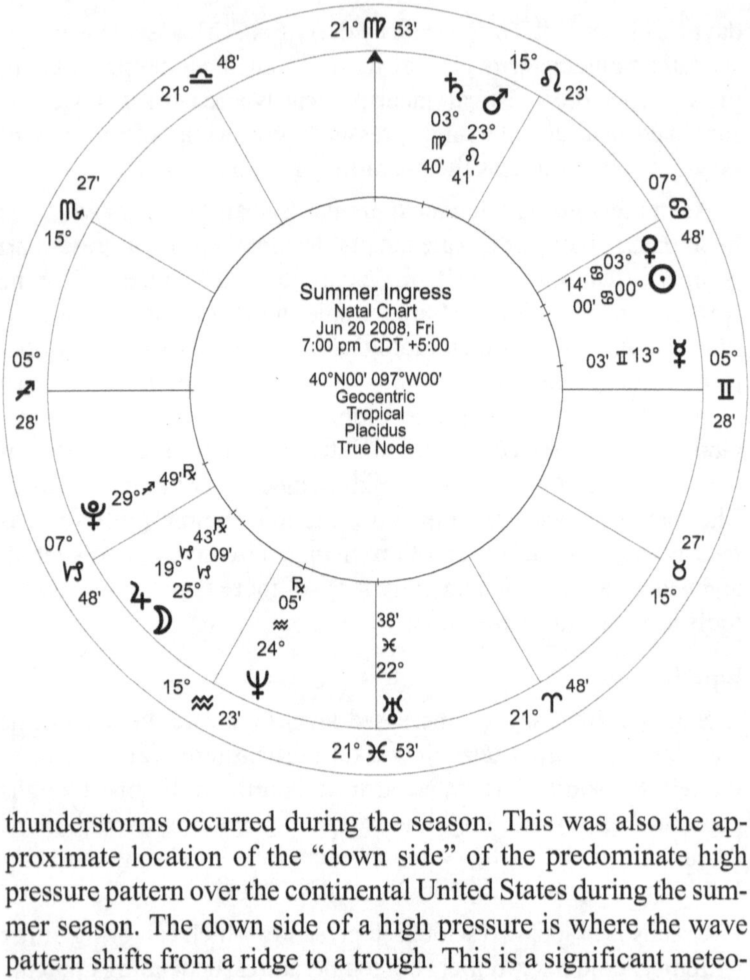

thunderstorms occurred during the season. This was also the approximate location of the "down side" of the predominate high pressure pattern over the continental United States during the summer season. The down side of a high pressure is where the wave pattern shifts from a ridge to a trough. This is a significant meteorological location, where air begins to gain the counterclockwise spin that reaches its maximum after it travels through the deepest part of the trough, and then heads upward toward another ridge.

This thunderstorm example (and the hurricane example below) suggests that Uranus also represents wind and conditions in the mid- and upper-troposphere. As warm, moist surface and low-level air rises in a thunderstorm updraft, it cools on ascent and then descends, helping to create the thunderstorm downdraft. With this

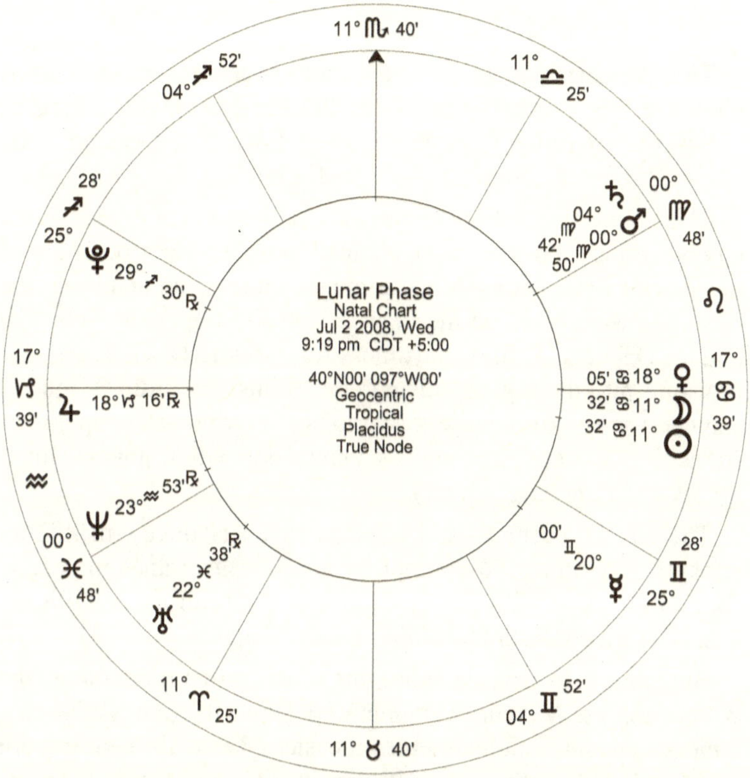

cold pool of air, represented by Uranus, thunderstorms would tend to be more severe.

The July 2, 2008 New Moon chart shows Mercury square lunar phase and ingress Uranus, which is a thunderstorm signature. Mercury (wind) also indicates a cold front (surface-based cold air which promotes low-level convergence), especially when in aspect to Uranus (upper-level divergence), so that week was favorable throughout that area for severe thunderstorm development. Moisture is indicated by Venus, which was in a separating trine to the lunar phase MC, an approaching trine to Uranus, and an opposition to Jupiter (heat). The lunar phase Sun-Moon conjunction in Cancer (moist sign) was trine-sextile the Scorpio-Taurus (moist signs) meridian, promoting the southerly flow of warm, moist air from the Gulf of Mexico that week.

Thunderstorms occurred in the multi-state area that week, including a July 8 squall (straight-line thunderstorms with damaging winds) that extended from southwestern Iowa to mid-eastern Wisconsin. It moved southeastward into Illinois.

This squall line had its genesis to the west much earlier in the day when fast mid-level winds helped move a weak trough out of the Rockies (transiting Mars conjunct ingress, lunar phase and transiting Saturn on the meridian in that location) into Nebraska and the Dakotas. At the same time, strong southerly winds brought moisture into the region, raising dew points (transiting Venus in Cancer trine-sextile ingress Meridian) as well as temperature (transiting Sun conjunct ingress Venus, opposition ingress, lunar phase, and transiting Jupiter).

The squall reached—and subsequently sustained—maximum strength, including 60-80 mph winds, as the transiting Moon formed a square with ingress Venus and a semisextile with the Mars-Saturn configuration (high winds, trough).

But what is so striking about this squall is that it paralleled the path of ingress Mercury conjunct-opposition the ingress Descendant-Ascendant which arced from eastern Texas through eastern Iowa and into northeastern Wisconsin. Of course, this Mercury line had to be activated, which it was at the lunar phase, when Sun, Moon, Venus, and Jupiter followed the same arcing path just west of the ingress Mercury line, where the path of these four planets intersected ingress, lunar phase, and transiting Uranus. This combination of Mercury and Venus resulted in a stationary front (a combination of cold and warm fronts) along the Mercury line and a low pressure where the planets intersected Uranus. This is an example of how meteorology and astrometeorology could be used to identify areas at high risk for severe thunderstorms and even better fulfill the Storm Prediction Center's mission to forewarn and protect the public. All the planetary alignments and meteorological factors were thus not only present for severe thunderstorms that week but also for the squall to develop July 7.

Using the ingress and lunar phase charts and that week's transits, an astrometeorologist would see the severe thunderstorm potential of Mercury square Uranus, and the heat and moisture from the Sun-Venus-Jupiter alignment, in addition to the moist signs on the lunar phase MC-IC. But many would put the timing a day earlier when the Moon, in Virgo, would have triggered the Mercury-Uranus square. However, by using the tools of meteorology, the astrometeorologist would be alerted to the eastward moving trough and front (originally a cold front that evolved into a stationary front) and thus that the area was at high risk on July 7 for severe thunderstorms. Then the arcing Ascendant-Descendant "lines" would take on new meaning, especially when viewed in context with the convergence-divergence pattern related to Mercury and Uranus, and the strong southerly flow of warm, moist air.

Hurricane Gustav.

Uranus also had a role in Hurricane Gustav, which made landfall in Terrebone Bay, Louisiana, September 1, 2008. The same solar ingress is valid for this hurricane.

The solar ingress chart has a hurricane signature: Mars opposition Neptune. An astrometeorologist would thus immediately know that summer 2008 had the potential to be an active tropical storm season. But because these planets aligned with the ingress meridian on the west coast, any hurricanes were unlikely to be Category 4 or 5. Had thse planets been conjunct or square the meridian in an eastern hurricane-prone area, the opposite would be true. Severe thunderstorms, however, would occur more often than usual in the Gulf states, where Uranus was influential. The Sun-Pluto opposition square the ingress meridian in Louisiana, also increased the potential for high winds, whether from thunderstorms or tropical storms.

Neptune represents moisture and cyclonic circulation, a characteristic of tropical storms, and is often a factor, along with Saturn, in significant low-pressure systems (another tropical storm characteristic). Mars reflects the heat necessary for tropical storms to develop.

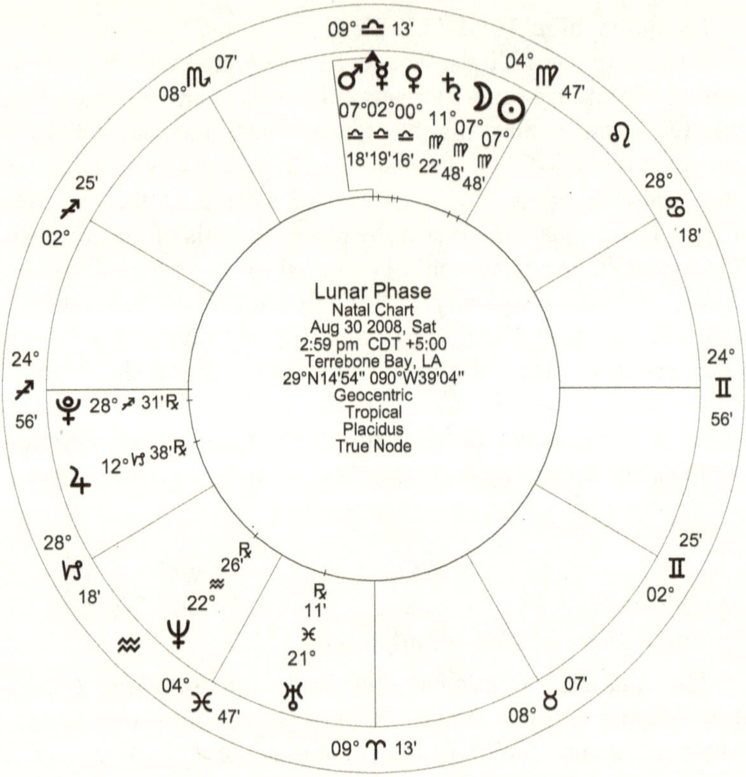

Astrometeorology can be used to identify any week (lunar phase-solar ingress comparison) when an area or region might be prone to a tropical storm. But its real value is in pinpointing the exact longitude and latitude of landfall once the many meteorology tools (mostly satellite- and aircraft-based) available to tropical forecasters have identified a tropical storm and its projected path.

Mars in the August 30 New Moon chart activated the solar ingress Mars-Neptune opposition through a semisquare. In itself, this was not enough to indicate a hurricane. Only the possibility of a tropical storm in the area could be forecast. Nor was lunar phase Venus conjunct-opposition the ingress MC-IC and square the ingress Sun-Pluto opposition. But this alignment would definitely alert the astrometeorologist to severe weather during that week. In addition, transiting Mars would cross the 90W Midheaven within two days of the lunar phase.

Ingress Uranus (synoptic) was likely the key planet related to Gustav's weakening from a Category 3 to a Category 1 hurricane between the time it entered the Gulf and made landfall, and as it moved inland. Its ingress placement had set the trend for the season. This cold, dry planet appears to be related to mid- and upper-troposphere winds, and in fact, meteorology shows that Gustav weakened because of cold, dry steering winds that diminished the warm, moist conditions any hurricane needs to survive. On September 1, there was a ridge of high pressure (Uranus) almost directly north of Louisiana, from Missouri to Canada.

At the time of landfall, transiting Venus was conjunct lunar phase Mercury, and both transiting Venus and Mercury were square the ingress Sun-Pluto opposition. This was triggered by the transiting Moon, which crossed the ingress Midheaven and formed a conjunction with ingress Venus to square the ingress Sun-Pluto opposition. But what exactly pinpoints Terrebone Bay, Louisiana, as the location, and 9:00 a.m. as the time of landfall is the alignment of the transiting Midheaven with the lunar phase Midheaven, both of which (along with Mars) aspected the ingress Mars-Neptune opposition by semisquare-sesquisquare.

The use of both meteorology and astrometeorology could thus give the public early warning of where a hurricane would make landfall—as much as a week or more in advance instead of a few days at most.

Much research still needs to be undertaken in order to better correlate the tools of meteorology and astrometeorology as well as how the planets, signs, and planetary alignments represent the various atmospheric factors in weather development. The results could potentially broaden and deepen both fields of study with the ultimate goal of safeguarding lives and providing economically advantageous information.

60-YEAR INTERVALS AND OCTOBER PANICS

By David McMinn

Abstract: A survey of Financial Panics occurring in October shows that they tend to occur at 60-year intervals. A list of those that have occurred in the 19th and 20th centuries is given.

A key interval of 60 years was found to be strongly associated with October panics (McMinn, 2006, 2009) over recent centuries. Adding or subtracting 60 years from the dates of major October DJIA annual one day falls consistently produced corresponding annual one falls between August 19 and December 20. This finding was repeatable for the corresponding DJIA annual one day rises. The 60 year intervals also showed up for 19th century US October panics. Additionally, 6 October AOD falls (=> -3.60%) occurred in the period 1910 to 2000 and they formed a very neat grid based on 2-8/60 years. The 6 events occurred before the full Moon and before the new Moon and produced a precise pattern that would be very unlikely to occur by chance.

Why the interval of 60 years was so important in the timing of October panics remains unknown. It probably has to do with Moon Sun tidal effects, but otherwise the problem remains very enigmatic.

Background.

The annual one day (AOD) fall is the biggest percentage one day fall in the year beginning March 1, where as an AOD rise is the biggest corresponding percentage rise. The time assessed was 12:00 p.m. EST on the day of the panic or AOD fall, with no adjustment being made for daylight saving time. E° is the abbreviation used to denote the degree on the ecliptical circle, where as A° is used for the angular degree between the Moon and Sun (lunar phase). This was to prevent confusion between two very different concepts. For the period 1885 to 1896, the 14 Stock Average and the 12 Stock Average indexes were used. Post 1896, the Dow Jones Industrial Average was utilized.

October DJIA AOD Rises and Falls

Appendix 1 presents all 10 major DJIA AOD falls (=> -3.60%) occurring between September 10 and October 31 since 1885. By adding or subtracting 60 years, each of these autumn events had a corresponding AOD fall (=> -2.45%) between August 19 and December 20, with NO EXCEPTIONS. DJIA AOD falls late in 2015 and 2046 could be expected by extrapolating from this pattern.

60 year intervals were also evident for major DJIA AOD rises (=> +4.00%) occurring between September 24 and November 5 (see Appendix 2). Most of these rises had a corresponding DJIA AOD rise (=> +2.50%), plus or minus 60 years between August 20 and December 30 (see Appendix 2). However, some of these AOD rises were quite small at between +2.50% and +3.10% as in 1897, 1918, 1927 and 1991. The anomalies happened in 1914 when the New York stock market was closed and in 1948 when no AOD rise over +2.10% was recorded.

In 1914, there was a complete lack of data between July 30 and December 12, when the US stock market was closed following the outbreak of WW I. Presumably, a security rally would have most likely taken place, once the US government confirmed that it would remain neutral in the conflict. This occurred on September 7, 1939 when President Roosevelt announced that the USA would not be entering WW II.

A summary can be given of the findings.

- AOD Falls (=> -3.60%) between September 10 and October 31.
- Time Range All AOD Falls in Appendix 1, August 19-December 20

- AOD Rises (=> +4.00%) between September 24 and November 5.
- Time Range All AOD Rises in Appendix 2, August 20-December 30

Historical October Panics

60 year intervals were also important in the timing of 19th century US panics occurring between September 20 and October 31:

- 1839, US panic (Oct) +60 1899 DJIA AOD fall (December 18).
- 1847, British panic (October 23) +60 1907 US banking panic (October 22).
- 1857, US & British panics (October 14) +60 1917 DJIA AOD falls (November 1 and 8).
- 1869, US Black Friday (September 24) +60 1929 US Black Tuesday (October 29).
- 1871, US Chicago fire panic (October 09) +60 1931 DJIA AOD fall (Sepember 24).
- The 1907 banking panic (October 22) was anomalous, as 1967 had no AOD rise or fall over 2.00%.
- 1873, US Black Friday (September 19) plus 60 years gave the 1933 AOD fall of July 21. This was a little out compared with the AOD falls in Appendix 1, which took place from August 19 to December 20.

DJIA October AOD Falls 1910-2005

Between 1910 and 2005, DJIA October AOD falls (=> -3.60%) were experienced in 1927, 1929, 1937, 1987, 1989 & 1997. They yielded a precise grid based on 2-8/60 solar years (see Appendix 3), in which lunar phase took place within very narrow ranges 150-165 A° (before a full Moon) and 320-330 A° (before a new Moon). Most DJIA AOD October falls were followed a few days later by an AOD rise, with the September 6, 1927 AOD rise being the only irregularity.

The 2-8/60 year grid may be extended to the right hand side by adding 11 years. This gave the October 15, 2008 AOD fall which happened during Black October. Subtracting 60 years from this date gave the November 3, 1948 AOD fall (-3.85%), a near panic induced by Truman's surprise victory in the presidential elections.

Moon Sun Cycles

Major October DJIA AOD falls (=> -3.60%) in Appendix 1 consistently had lunar phase around the full Moon (150-205 A°) and before the new Moon (300-350 A°). This also applied to the early historical October panics in 1847, 1857, 1871 and 1907 (no precise date was available for the 1839 panic). The non-October events showed no such restricted distributions for lunar phase.

For all 12 October DJIA rises in Appendix 2, lunar phase always occurred around the full Moon (160-200 A°) and after the 3rd quarter Moon to after the new Moon (280-015 A°). This gave a similar distribution to the major AOD falls in Appendix 1.

October Events	*No*	*Lunar Range*	*No*	*Lunar Range*
AOD Falls (=> -3.60%)	4	150-205 A°	5	300-350 A°
AOD Rises (=> +2.50%)	3	160-200 A°	9	280-015 A°

No established Moon Sun cycle could be linked to the key interval of 60 years. McMinn (2009) proposed 30 and 60 year cycles based on the cycles of the Metonic (235 lunar months) and Octaeteris (99 lunar months), but this was pure speculation and was not supported by any evidence.

Conclusions

The interval of 60 years can be linked strongly with October panics over the past 200 years. Unusually, it was not associated with other panics and AOD falls occurring at other times of the year. The term 'interval' was used because the impact of 60 years was so variable, whereas the term 'cycle' would imply something with far more regular timing.

No explanation can be given as to why the interval of 60 years is so crucial in the timing of October panics. Presumably, it has something to do with Moon Sun tidal effects, but that is all that can be stated. The findings do not support the random walk view of financial activity, as promoted in traditional economic theory. Rather the timing of major shifts in financial activity is mathematically structured and quite precise.

From **McMinn, David**. *Market Timing Moon Sun Research 2006-2009*. Privately published. 2009.

References

McMinn, David. *Market Timing By The Moon & The Sun*. Privately published. 2006.

McMinn, David. *Market Timing Moon Sun Research 2006-2009*. Privately published 2009.

Appendix 1, 60-Year Intervals and Autumn DJIA AOD Falls Over -3.60%

Time Spans: 1895-1926 and 1955-1986

1895		1897		1903		1926	
Dec 20		Sep 21		Aug 19		Oct 15	
-6.61%		-3.95%		-4.07%		-2.73%	
051 A°	+2	299 A°	+6	336 A°	+23	104 A°	+1
		Oct 12		Oct 19			
		-3.90%		-4.17%			
		202 A°		348 A°			
	+60		+60		+60		+60
1955		1957		1963		1986	
Sep 26		Oct 21		Nov 22		Sep 11	
-6.54%	+2	-2.48%	+6	-2.89%	+23	-4.61%	+1
118 A°		339 A°		070 A°		095 A°	

Time Spans: 1926-1931 and 1986-1991

1927		1929		1931		1937	
Oct 8		Oct 29	+2	Sep 24		Oct 18	
-3.65%	+2	-11.73%		-7.07	+6	-7.75%	+4
150 A°		326 A°		157 A°		164 A°	

	+60		+60		+60		+60
1987		1989		1991		1997	
Oct 19		Oct 13		Nov 15		Oct 27	
-22.61%	+2	-691%	+2	-3.93%	+6	-7.18%	+4
324 A°		164 Ao		102 A°		320 A°	

Time Spans: 1931-1948 and 1991-2008

1941		1948		1955		1986
Dec 8		Nov 3		Sep 26		Sep 11
-3.50%	+7	-3.85%	+7	-6.54%	+31	-4.61%
234A°		032 A°		118 A°		095 A°
+60		+60		+60		+60
2001		2008		2015		2046
Sep 11		Oct 15		?		?
na	+7	-7.75%	+7		+31	
281 A°		192 A°				

[1]Due to the outbreak of WW I, the New York stock market was closed on July 30 and did not reopen until December 12. Thus no conclusions may be drawn.

Appendix 2, 60-Year Intervals and Autumn DJIA AOD Rises Over +4.00%

Time Spans: 1897-1918 and 1957-1978

1897		1903		1914[1]		1918	
Aug 31		Oct 16					
+2.97%	+6	+5.11%	+11		+4	+2.67%	+9
314 A°		309 Ao				315 A°	
+60		+60		+60		+60	
1957		1963		1974		1978	
Oct 23		Nov 26		Oct 9		Oct 20	
+4.12%	+6	+4.50%	+11	+4.71%	+4	+4.46%	+9
007 A°		120 A°		282 A°		011 A°	

Time Spans: 1927-1937 and 1987-1997

1927		1929		1931		1937	
Sep 6		Oct 30		Oct 6		Oct 20	
+2.95%	+2	+12.34%	+2	+14.87%	+6	+6.07%	+4
116 A°		338 A°		294 A°		191 A°	
+60		+60		+60		+60	
1987		1989		1991		1997	
Oct 21		Oct 16		Aug 21		Oct 28	
+10.17%		+3.43%		+3.02%		+4.71%	
347 A°	+2	197 A°	+2	139 A°	+6	330 Ao	+4
				Nov 23			
				+3.00%			
				204A°			

Time Spans: 1941-1948 and 2001-2008

1941		1948[1]		1957		1974	
Dec 30				Oct 23		Oct 9	
+3.50%	+7		+9	+4.12%	+17	+4.17%	
167 A°				007 A°		282 A°	
+60		+60		+ 60		+ 60	
2001		2008		2017		2034	
Sep 24		Oct 13		?		?	
+4.47%	+7	+11.08%	+9		+17		
093 A°		165 A°					

[1] No DJIA AOD rise over +2.10% was recorded during the year.

Appendix 3, October AOD Rises/Falls and the 2-8/60 Year Grid

Lunar Month Intervals and Lunar Phase AOD Falls

1927		1929		1937
Oct 8		Oct 29[1]		Oct 18
-3.65%	+25.47	-11.73%	+98.58	-7.75%
150 A°		326 A°		164 A°
+742.48		+741.57		+742.41
1987		1989		1997
Oct 19		Oct 13		Oct 27
-22.61%	+24.55	-6.91%	+99.42	-7.18%
324 A°		164 A°		320 A°

Solar Year Intervals AOD Fall

1927		**1929**		1937
Oct 8	+2.06	Oct 29[2]	+7.97	Oct 18
+60.03		+59.96		+60.03
1987		**1989**		**1997**
Oct 19	+1.98	Oct 13	+ 8.04	Oct 27

AOD Rises

1927	1929	1937
Sep 6	Oct 30	Oct 20
+2.95%	+12.34%	+6.07%
116 A°	338 A°	191 A°
1987	1989	1997
Oct 21	Oct 16	Oct 28
+10.17%	+3.43%	+4.71%
347 A°	197 A°	330 A°

[1]The 1929 AOD fall happened on Monday October 28, although the following day, Black Tuesday, has been used in the table.

[2]The 1929 AOD fall happened on Monday Oct 28, although the following day, Black Tuesday, has been used in the table.

Distant Planetoids

by James Herschel Holden, M.A. FAFA

Abstract: A discussion of some of the recent discoveries of sizable planetoids in the far-out Kuiper Belt region of the solar system.

During the last twenty years astronomers have made remarkable discoveries of planetoids in the far reaches of the solar system. The generic name for all of them is *Trans Neptunian Objects* or TNO's. The majority of those so-far discovered lie within the so-called Kuiper Belt—a region from about 30 astronomical units (au)[1] to 55 au. Planetoids at those distances are also called *Kuiper Belt Objects* or KBO's, and would have zodiacal periods of from 163 to 402 years. The *Dwarf Planet* Pluto has a mean distance of 39.46 au with a period of about 245.6 years. Perhaps as many as 200 of the newly discovered KBO's have mean distances between 39 and 40 au and were called *Plutinos* (more recently called *Plutoids*). The new name *Cubewano* has been assigned to the 700 or so planetoids in the range 42.8 au to 48 au. Planetoids with mean distances greater than 50 au are called "scattered disk objects" or SDO's.

So far astronomers have discovered about 1300 TNO's, and they estimate that these are only a small fraction of an estimated 5,000 or more planetoids in the Kuiper Belt and beyond. We might reasonably suppose that astronomers may discover another few thousands in the next decade; but due to their visual faintness, the discovery of as many as 5,000 is probably not possible with existing telescopic equipment.

---

[1]The astronomical unit (au) is a common measure of distance used by astronomers. It is the mean distance of the Earth from the Sun. The Earth's mean distance is therefore 1 au.

As astrologers, we are concerned with the possibility that there may be discernible astrological influences associated with some of these far-out planetoids. And perhaps we might even wonder if any of those discovered so far correspond to the first four of the "Trans Neptunian Planets" postulated by Alfred Witte of the Hamburg School and used today by Uranian astrologers.

Of those, Cupido with a period of 262.5 years and Hades with a period of 360.7 years could be Kuiper Belt Objects (KBO's). Zeus with a period of 456 years and Kronos with a period of 522 years are outside the Kuiper Belt. It would be possible to scan the orbital elements of the KBO's so-far discovered and see if any of them correspond to Cupido or Hades. I have not done this.

And there are another two dozen or so hypothetical Planets postulated by other astrologers during the 20th century that could also be looked for.

Aside from all that, the method currently in vogue of assigning astrological influences to planetoids is to consider the name that is assigned to the planetoid. Unfortunately, only a handful of the 1300 KBO's have been assigned names. And the probability of names being assigned at any time soon to the others seems unlikely. (For some reason, the International Astronomical Union assigns names very slowly.) I have also seen estimates that there may be more than 100,000 asteroids in the classical belt between Mars and Jupiter. Many thousands of them have already been named, but naming all of them will be another major challenge for the astronomers.

How then should we consider this great wealth of planetoids that the astronomers have discovered and are expected to discover in the future? One method would be to investigate the largest ones on the assumption that they might have the greatest astrological influence. In the case of the distant asteroids, the latest list in order of size that I have seen is this: Eris, Pluto, Makemake, Haumea, Sedna, Orcus, Quaoar, Varuna, and Ixion.

First we should notice that Eris is somewhat larger than Pluto! This immediately brings up the question, Should it replace Pluto as the ruler of Scorpio? And if not, why not? Some astrologers might wish to assign it to the rulership of Libra or Virgo, but that seems inappropriate, since it is probably only a matter of time until the astronomers discover several more TNO's that are equally large, and what signs could they be assigned to?

And then there is Sedna! It is much farther out than the KBO's and therefore much slower in motion. It is estimated to have a zodiacal period of about 8,100 years! So far, it is the slowest moving planetoid yet discovered. It might be found to correlate in some way with historical periods.

Each of the TNO's above is described in more detail as follows.

Dwarf Planet 136199 Eris

This is the largest TNO discovered as of the end of 2007. It is somewhat larger (1660 miles in diameter) and more massive than the Dwarf Planet Pluto! Its discovery was announced by astronomer Mike Brown and his associates on 29 July 2005. It was later discovered to have a satellite that has been named Dysnomia. It has a very high albedo of 0.86 and is almost grey in color.

Eris has a far-out mean distance of 67.7 au, a zodiacal period of about 545 years, a large eccentricity of 0.4418, and a very high inclination to the ecliptic of 44.19°.

This planetoid was originally thought of by the discoverers as "Planet X" and was therefore named "Xena" after the name of the leading character in the popular TV series "Xena: Warrior Princess." It was later given the permanent official name Eris.

Eris was the Greek goddess of Discord, and her daughter Dysnomia was the goddess of Lawlessness (a punning reference to the name of the actress Lucy Lawless, who played the part of Xena's companion in the TV series).

It is currently near 21 ♈ and 14 S.

Dwarf Planet 134340 Pluto

Pluto was discovered by Clyde Tombaugh at Lowell Observatory at 4 PM on 18 February 1930. It was immediately hailed as the "Ninth Planet," a distinction that it held until the summer of 2006, when a small group of astronomers demoted it to the status of "Dwarf Planet." Since its demotion, it has been assigned the planetoid No. 134340.

The Planet was named for the Roman god of the underworld Pluto—a name that had the distinction of beginning with the letters P and L, which were the initials of Percival Lowell, who had instituted the search for it. Astrologers were delighted by the discovery. Some wanted to assign it to Aries, but the majority preferred Scorpio.

Its diameter is 1340 miles. Its mean distance is 39.46 au, its tropical period is 245.6 years, its eccentricity is a large 0.2459 and its inclination to the ecliptic is also a high 17.143°. It has three satellites. It is gravitationally coupled to Neptune by a 2:3 ratio.[1] Other TNO's in its vicinity that have that same gravitational coupling are called "Plutinos" (or, more recently, "Plutoids").

It is currently near 3 ♑ and 5N.

Dwarf Planet 136472 Makemake

Of the TNO's so far discovered, Makemake, with a diameter of 932 miles is thought to be slightly larger than Haumea,[2] and is thus the next smaller in size after the Dwarf Planet Pluto. So far, no satellites have been discovered. It was originally nicknamed "Easter Bunny" because it was discovered on 31 March 2005 near the date of Easter. It is red in color. Its mean distance is 45.79 au, its tropical period is 306.2 years, its eccentricity is 0.159, and its inclination to the ecliptic is a high 28.96°.

On 11 July 2008 the International Astronomical Union re-classified it as a Dwarf Planet and a Plutoid, and it was officially named Makemake (pronounced "mahkeh-mahkeh"). In the religion of the Rapanui people of Easter Island Makemake is the creator of humanity and the god of fertility.

It is currently near 27 ♍ and 29N.

---

[1]The ratio of periods is 2 X 245.58 = 491.16 and 3 X 163.723 = 491.17 – very close!

[2]However, most of the astronomical data on line seems to suggest that it is larger than Makemake.

Dwarf Planet 136108 Haumea

The next smaller planetoid in size after Makemake was originally nicknamed "Santa" because it was discovered a few days after Christmas on 28 December 2004. But on 17 September 2008 it received the designation of Dwarf Planet and the official name of "Haumea (pronounced "how-may-ah")." It is apparently a very rapidly rotating football-shaped object about 1200 miles long in its longest dimension. Its mean distance is 43.34 au, its solar period is 282.8 years, its eccentricity is 0.189, and its inclination to the ecliptic is a high 28.19°.

It was subsequently discovered to have two satellites that were nicknamed "Rudolph" and "Blitzen," but they are now officially named Hi'aka and Namaka. Haumea is the Hawaiian goddess of childbirth and fertility; Hi'aka and Namaka are two of her children.

It is currently near 17 ♎ and 27N.

TNO 90377 SEDNA[1]

The remarkable planetoid Sedna traverses the outermost regions of the solar system, sometimes called the "Inner Ort Cloud." It was discovered on 14 November 2003, and the discovery was announced to the public at 10 A.M. California time on 15 March 2004. Its great mean distance and eccentricity present a puzzle to astronomers.

Sedna's diameter is uncertain. The latest estimate that I have seen is 995 miles. The latest calculated orbit has it at a mean distance of 519.5 au, with a zodiacal period of 8,113 years, a very large eccentricity of 0.8536, and an inclination to the ecliptic of 11.932°. At its closest, it will still be 76 au from the Sun, while at its farthest distance it is 963 au from the Sun. However, its orbital elements are uncertain, because it has only been observed since 1990. It is said to be very red in color (the "second reddest body in the Solar system").

---

[1]A longer preliminary account was published in the A.F.A. periodical *Today's Astrologer*, Vol. 66, No. 3 (3-20-2004): 4-6.

It was subsequently named Sedna, the name of an ice-goddess of the Inuit people. That name was chosen because the planetoid's surface temperature must be close to absolute zero due to its great distance from the Sun.

At present, Sedna is about 88 Astronomical Units or approximately 8 billion miles from the Sun. This contrasts with Pluto's present distance of 30.8 au or 2.9 billion miles from the Sun. But Sedna will be in perihelion in 2076, when it will be about 76 au from the Sun.

Sedna's current position is near 21 Taurus and 12 South Latitude. It is currently moving about 0.56 degree per year. But its velocity will slowly increase to about 0.76 degree per year at perihelion, when it will be in about 8 Cancer.

Astrologers will naturally wonder what to make of this new astronomical discovery. It would seem best to treat Sedna as most astrologers have treated the previously discovered planetoids, namely as individual planetary bodies that may perhaps have some distinctive significance but no sign rulership.

However, unlike the closer-in planetoids, the very slow motion of Sedna (hardly more than 47 degrees in the next 72 years) would seem to require a different treatment of its astrological effects. Its position in secondary directions would be nearly fixed at its natal position. And its transits would hardly differ from one year to the next. Thus, to some extent, it might be considered to be a slowly moving "sensitive point" in the zodiac—somewhat like one of the still slower fixed stars.

At present, the elements of the orbit of Sedna are only approximate, and they will remain so for many years; consequently, positions calculated from them are increasingly uncertain for dates several centuries removed from the present. However, to give some idea of the great variability of Sedna's motion, I have estimated that it entered the sign Capricorn about 47 B.C., Aquarius about 1039 A.D., and Pisces about 1628 A.D.

Modern positions can be calculated with more certainty, and heliocentric longitudes and latitudes of Sedna are shown below for

January 1st at 10-year intervals in the following table. Geocentric longitudes would not differ by more than 0.8 of a degree one way or the other from the heliocentric longitudes, and the latitudes would differ by a negligible amount. More precise geocentric positions for particular dates can be obtained from one of the on-line ephemerides.

1900	8.0 ♈	-8.5	1960	27.2 ♈	-10.7	2020	27.5 ♉	-11.9
1910	10.6 ♈	-8.9	1970	1.3 ♉	-11.0	2030	4.0 ♊	-11.8
1920	13.5 ♈	-9.2	1980	5.8 ♉	-11.3	2040	10.8 ♊	-11.5
1930	16.5 ♈	-9.6	1990	10.7 ♉	-11.6	2050	17.9 ♊	-11.0
1940	19.8 ♈	-10.0	2000	15.9 ♉	-11.8	2060	25.2 ♊	-10.3
1950	23.4 ♈	-10.3	2010	21.5 ♉	-11.9	2070	2.7 ♋	- 9.5

The negative sign preceding the Latitude indicates it is South.

TNO 50000 QUAOAR

This planetoid was discovered on 4 June 2002. It is a Cubewano KBO and is described as being moderately red in color. It has a recently discovered satellite, which will enable its mass to be determined.

Quaoar's mean distance is 43.40 au, its zodiacal period is 283 years, its eccentricity is a very small 0.034, and its inclination to the ecliptic is 7.98°. Its diameter is about 708 miles or a little smaller than Pluto's largest moon Charon (diameter 753 miles). And is thus larger than the Dwarf Planet 1 Ceres, and it is much larger than the current astrological favorite 2060 Chiron (originally classified as an asteroid, but now considered to be a comet), whose diameter is estimated to be only around 110 miles.

Astronomers have adopted the custom of naming the far-out KBO's for deities related to the creation of the world (cf. 20000 Varuna, named for a Hindu deity) because the Kuiper Belt is a region in the outer reaches of the solar system that is thought to have played an important part in the creation of the solar system.

Quaoar (pronounced 'kwah-o-wahr') is the name of a divinity in the religion of the Tongva people (also called the San Gabrielino Native Americans) who inhabited the Los Angeles area. Quaoar is the great force of creation that sings and dances the high ones of the Tongva into existence, according to Mack Acuña, a Tongva scholar and elder.

It is currently near 21 ♐ and 7N

TNO 90482 ORCUS

The planetoid Orcus was discovered on 17 February 2004. It is a neutral colored Plutoid. According to one estimate, its diameter is about 565 miles, although other astronomers think that it is a little larger than Quaoar. Orcus has one satellite.

Its mean distance is 39.419 au, its solar period is 245 years (very nearly the same as Pluto), its eccentricity is 0.2255, and its inclination to the ecliptic is 20.55°

Orcus was a Roman god of the Underworld and a punisher of dead evil-doers.

It is currently near 0 ♍ and 18S.

TNO 20000 VARUNA

This TNO was discovered on 28 November 2000. It is a Cubewano KBO of about 500 miles in diameter, and is moderately red in color. Its mean distance is 43.13 au, its zodiacal period is 280 years, its eccentricity is a very small 0.051, and its inclination to the ecliptic is 17.2°.

This planetoid was named after the ancient Hindu deity Varuna, who was a supreme divinity in the Vedas. He was the creator of the universe and the guardian of Cosmic law. One of his functions was the judgment of men and the punishment of sinners. However, his attributes and functions were assigned one by one to other deities as the centuries passed, so that finally he was merely regarded as the regent of the western quarter of the world and the Lord of the Ocean. In the latter respect he has some likeness to the Western god Poseidon or Neptune, all the more so since a thousand white horses form part of his retinue, and Poseidon was also said to rule horses.

It is currently near 21 ♋ and 4N.

TNO 28978 Ixion

Ixion was discovered on 22 May 2001. It is a Plutoid and is moderately red in color. Its diameter is a little more than 450 miles, and its mean distance is 39.68 au, with a zodiacal period of 247.55 years, an eccentricity of 0.242, and an inclination to the ecliptic of 19.58°.

In Greek mythology, Ixion was an evil-doer, the first person to commit the terrible crime of slaying a kinsman. For this, Zeus had him bound to a fiery revolving wheel as an eternal punishment.

It is currently near 16 ♐ and 2S.

ASTROLOGICAL RESEARCH

Eleven Police Officers Killed While on Duty.

By James Herschel Holden, M.A., FAFA

Abstract: A statistical discussion of the astrological birth data of a group of police officers who were shot dead while performing their duties.

During a 20-year period (1985-2005), 11 Arizona Police Officers were shot dead while trying to make an arrest. Their names and the dates of their demise were published in a table in *The Arizona Republic* in May 2005. I was able to obtain the birthdates of all eleven from the SSDI, and I took the state where their Social Security cards were issued to be their birth states. None of the birthdates was timed. I used noon in their assumed birth state as the nominal birthtime in each case. Their names, birthdates, state of birth, death date, and state of death are shown in the following table.

Name	*Birthdate*	*State*	*Death Date*	*State*
Crowe, Michael L.	07.10.1953	CA	04.07.1995	AZ
Martin, Robert K.	11.08.1938	AZ	15.08.1995	AZ
Blair, Kenneth R.	09.12.1948	NM	28.09.1995	AZ
Atkinson, Marc T.	16.08.1970	UT	26.03.1999	AZ
Moritz, Jeffrey A.	23.12.1969	AZ	21.06.2000	AZ
Lopez, Jason P.	28.12.1974	AZ	04.05.2001	AZ
Mauldin, Donald L.	07.10.1942	TX	24.08.2002	AZ
Hardesty, Patrick K.	16.03.1963	NM	26.05.2003	AZ
White, Eric J.	24.07.1974	IL	28.08.2004	AZ
Wolfe, Jason A.	13.12.1976	AZ	28.08.2004	AZ
Uribe, David C.	06.02.1957	AZ	10.05.2005	AZ

For the 11 names, I summarized their aspects and planetary sign positions; and I created astronomical expectancy files for planets in signs and for the planetary aspects with a 6° orb using proprietary software that I developed in the early 1980s.[1] The significant deviations from expectancy and their probability[2] (according to the binomial theory) were as follows:

Mercury right semi-sextile Mars	0.001
Sun in Leo	0.062
Mars right trine Saturn	0.001
Mercury Direct in Capricorn	0.009
Sun conjunct Mars	0.003
Venus Direct in Libra	0.036
Sun right trine Saturn	0.003
Mars Direct in Leo	0.081
Moon opposite Sun	0.006
Moon right trine Mars	0.051

Since the charts were untimed, the Moon aspects and their probabilities are subject to some error. But the number of planetary sign positions and interplanetary aspects, and their respective probabilities should be approximately correct.

These officers had several things in common. First, they were

[1]I originally developed the software in 1978 to run on a main frame computer. Interplanetary aspects and planetary sign positions were calculated for the years 1800-2000, and the tables for the years 1900-1949 were published as *Relative Frequency Tables* (Tempe, Az.: A.F.A., Inc., 1978). A few years later, when desktop computers became available, I modified the software to run on them. A file containing the birthdates is created and used as input for a program that calculates the planetary positions and mutual aspects and outputs files containing the observed sign positions and aspects by years. Companion programs calculate the astronomical expectancies for sign positions and aspects for all the years. Then, by comparing the observed positions and aspects with the expectancies, the significant deviations from expectancy can be noted and their probabilities calculated.

[2]Probabilities < 0.05 are generally considered to be significant. The smaller the number, the more significant the indication.

police officers. Second, they worked in a single location. And third, they died violently while on duty. Some of the aspects and planetary positions may therefore be characteristic of their profession and common location. But Sun conjunct Mars must certainly be a possible indicator of a violent death. The Sun afflicted by Mars is frequently found in the case of persons who have died violently from various causes, including gunshot and auto accidents. And the aspects of the Sun and Mars to Saturn and the Moon to Saturn are also significant due to the conjunction of the Sun and Mars.

The weekday distribution showed 4 officers born on Wednesday, 2 born on Thursday and Saturday, and 1 born on Monday, Tuesday, and Sunday; but no one was born on Friday. These results were not quite statistically significant (4 born on the same weekday having a probability of 0.060.)

Further research could be undertaken to examine their secondary progressions and transits at death.

WHAT WILL HAPPEN NEXT?[1]

by James Herschel Holden, M.A., FAFA

Abstract: A short history of the development of Western Astrology and the techniques employed in it to predict the future from the time of the Babylonians down to the present day, and an account of some rulers who employed astrological consultants.

Nearly every human being at one time or another has asked himself, *What will happen next?* His motive for asking that question may range from idle curiosity to sheer desperation. This innate desire has driven man to attempt to determine the future. Many different methods of doing this have been tried, and some are still practiced. One of the oldest methods that is still in common use is astrology, which presupposes a connection between appearances in the sky and occurrences on the earth.

The basic premise of astrology is that time itself, like the weather, has characteristics that change in quality from moment to moment. It is not uniform and neutral. The corollary is that anything that happens at a particular moment is influenced in some determinable way by the characteristics of that moment in time. Hence, a situation that occurs, a birth, a question that is asked, or an action that is initiated is influenced by those characteristics. And a further corollary is that those characteristics are recorded in some manner.

The characteristics of the sky at a particular moment in time re-

[1]This paper was read on 26 July 2004 at the Conference "Horoscopes and History" that was held by the Philosophy Department of the University of Amsterdam. An Italian translation was subsequently published in the quarterly *Linguaggio Astrale* No. 140 of the Centro Italiano di Astrologia (CIDA) on pp. 3-23.

flect the combination of influences arising from the signs of the zodiac, the location of the planets in the zodiac, the angular relationships of the planets, and the orientation of the zodiac with respect to the horizon. These characteristics as a whole can be conveniently referred to by using the Latin term for the sky, *Caelum*.

The basic idea of astrology was conceived by the Babylonians in the fifth century B.C. They subsequently invented the 360-degree circle and devised a fixed or "sidereal" zodiac of twelve 30-degree *signs*. Then, by constructing a star catalogue, they were enabled to reduce several centuries of observations into ingenious step-by-step tables of the motions of the Sun, Moon, and planets.[1] With this information available, they then postulated significations for the positions of the celestial bodies in the signs of the zodiac. They also took special notice of the celestial circumstances of eclipses of the Sun and Moon.

The Egyptians had no astrology, but they had devised a ring of 36 asterisms that were later called *decans*. These were used to tell time at night, and since they had individual names, they were considered to be either "lucky" or "unlucky." Although there is no record of it, it seems that the idea occurred to them that the asterism on the eastern horizon at the time a child was born gave an indication of whether its life would be "lucky" or "unlucky."

In the second century B.C., the city of Alexandria, Egypt, had a mixed population of Egyptians, Greeks, and Jews. Some persons whose names are unknown combined the astrological ideas of the Babylonians that were known to the Jews with the Egyptian assessment of luck from the rising asterism and with Greek philosophical and religious concepts. From that amalgam they invented Horoscopic Astrology. They wrote treatises defining the rules of the new art, and they prepared a set of astronomical tables referred to a fixed zodiac.[2] These treatises were issued under the names of

---

[1]O. Neugebauer, *Astronomical Cuneiform Texts*, 3 vols. (London: Lund Humphries, 1955.).

[2]The so-called *kanones aiônies* or *Eternal Tables*. See Note 1 on p. 93 below.

Nechepso, Petosiris, Hermes, Abram, and Aesculapius. Unfortunately, none of them have come down to us intact, but only partially in citations by later astrologers. And from the outset, Greek was the principal language of astrology.

The basic features of Horoscopic Astrology are these: the ecliptic is divided into the twelve signs of the zodiac. Each of these signs has a distinctive character. They are classified in groups of two, of three, and of four. Thus, the signs beginning with Aries are alternately male and female. And also beginning with Aries, they are considered to have an elemental quality of Fire, Earth, Air, or Water. Finally, still beginning with Aries, they have one of three characteristics: Cardinal, Fixed, or Mutable. These classifications, together with their rulership by the planets, establish the fundamental natures of the signs.

The planetary rulerships of the signs are assigned in the order of the velocity of the planets measured in both directions from the two signs Cancer, which is ruled by the Moon, and Leo which is ruled by the Sun. Thus, Gemini and Virgo are ruled by Mercury, Taurus and Libra by Venus, Aries and Scorpio by Mars, Pisces and Sagittarius by Jupiter, and Aquarius and Capricorn by Saturn. (Modern astrologers have altered the rulerships to make Uranus ruler of Aquarius, Neptune ruler of Pisces, and Pluto ruler of Scorpio.) The planets were also divided into sects: Saturn, Jupiter, and the Sun belonged to the diurnal sect, Mars, Venus, and the Moon to the nocturnal sect, while Mercury was variable.

If a planet chances to be in its own sign, its nature is strengthened, but if it is in a sign opposite to the one that it rules, its nature is thwarted to some extent. It was also found (by the Babylonians) that certain signs enhance the action of the planets. These were later called *Exaltations*; and the signs opposite those, that diminish the action of the planets, were called *Falls*. The Alexandrians also defined some other minor rulerships, in particular the dodecatemories (the twelfths of the signs, devised earlier by the Babylonians), the decans, and the five-fold subdivision of the signs into "terms," but those mentioned previously are the principal ones.

Since all the celestial bodies are in motion, each day presents a unique pattern in the zodiac and thereby defines the characteristics of time itself at a particular moment.

The orientation of the zodiacal signs is also important. The one on the eastern horizon is called the Ascendant. And counting counter-clockwise, they constitute the celestial houses, which are called: Ascendant, second, third, IMC, fifth, sixth, Descendant, eighth, ninth, Mid Heaven, eleventh, and twelfth. The 1st, 4th, 7th, and 10th houses have the most power and are called *angular*; the 2nd, 5th, 8th, and 11th are of medium power and are called *succedent*; the remaining three, the 3rd, 6th, 9th, and 12th have the least power and are called *cadent*. Each of these houses was determined to preside over a certain feature of human life: the 1st house over personality, the 2nd house over finances, the 3rd house over brothers and sisters, the 4th house over parents, the 5th house over children, etc.

Astrologers also consider the approximate angular distances between the planets to be *aspects* and call the near zero distance *conjunction*, the near 60° distance *sextile*, the near 90° distance *square*, the near 120° distance *trine*, and the near 180° distance *opposition*. A variation of a few degrees from exactitude is allowed in each case and is called *orb*. Of these aspects, the conjunction is neutral, the sextile and trine are harmonious, and the square and opposition are disharmonious.

One other notable invention of the Alexandrian astrologers was the assignment of planetary rulers to the individual days of the seven-day Jewish week. They began by assigning Saturn as the ruler of the Jewish Sabbath and the ruler of the first hour of that day. Then the succeeding hours were assigned in the *heptazone* order (Saturn, Jupiter, Mars, Sun, Venus, Mercury, Moon). And since there are 3 sets of 7 hours plus 3 left over in each 24-hour period, the ruler of the first hour of the next day was 3 steps down in the series, or the Sun. Hence the next day was Sun-day, the one after that Moon-day, etc.

Although this is nowhere explicitly attested as being an invention of the Alexandrian astrologers, it must have been. The first explicit mention of a named weekday is in Book I. 3. 18 of the poems of Albius Tibullus (c.54-19 B.C.),[1] where he has the phrase *Saturni sacra dies* 'the day sacred to Saturn'. The fourth century astrologer Paul of Alexandria devoted two chapters of his book *Introduction* to explain how to determine the day of the week and the planetary ruler of each hour.[2]

This astrological invention eventually replaced the 10-day week of the Egyptians, the 8-day week of the Romans, and gave names to all the days of the Jewish 7-day week. It has endured for two millennia and has come into world-wide use—a noteworthy example of the influence of astrology on everyday life.

Finally, we may mention the Lots. These were perhaps devised by the astrologer who wrote under the name Hermes. They are points in the zodiac (originally houses in the horoscope chart) that are determined by finding the distance between two points and adding that distance to the ASC or the ASC degree. They were seven in number at first[3]—the two most famous being the Lot of Fortune (now called the Part of Fortune), which was the distance from the Sun to the Moon added to the ASC. The second (the Lot of the Spirit) was the reverse: the distance from the Moon to the Sun added to the ASC. And in both cases, the order of the planets was reversed for a nocturnal nativity.

Many other Lots were devised by Hermes's successors (Firmicus mentions more than twenty, calling them "houses"),[4]

---

[1]See F.K. Ginzel, *Handbuch der mathematischen und technischen Chronologie*, 3 vols. (Leipzig: J.C Hinrichs, 1914.), vol. 3, p. 98.

[2]Paul of Alexandria, *Introduction* ed. by Emilie Boer (Leipzig: B.G. Teubner, 1958.), Chapters 20-21, certainly based on much older astrological texts.

[3]The original Greek text is preserved in Paŭl of Alexandria, Chapter 23 "The Seven Lots of those in the *Panaretos*." These and many others are mentioned in several lists given in Heliodorus, *Commentary on Paul* ed. by Emilie Boer (Leipzig: B.G. Teubner, 1962.), Chapter 22.

[4]Julius Firmicus Maternus, *Mathesis*, 2 vols. (Leipzig: B.G. Teubner,

but Ptolemy muddied the water by only mentioning the Lot of Fortune and giving the same rule for both day and night nativities. These Lots (or Parts as they are now usually called) grew in popularity during late antiquity and the Middle Ages,[1] but since Ptolemy had only mentioned the Part of Fortune, the others were largely ignored by modern astrologers. A recent book by Zoller has re-aroused interest in them.[2]

Astrology itself is divided into four categories: Mundane, Natal, Horary, and Elections.

Mundane Astrology strives to analyze the celestial conditions that exist in the *Caelum* at a particular time and a particular place in order to determine their current and future influence on that particular location and the living things in it.

Natal Astrology strives to analyze the celestial conditions at the moment that a child is born in a particular place and at a particular time. It determines the influence of the *Caelum* at the time of birth and the subsequent influences that will occur later during the lifetime of that individual.

Horary Astrology strives to analyze the *Caelum* at the moment that a question is propounded at a particular place and time. From the analysis of a chart erected then, it proposes to determine the circumstances attendant upon that question and those affecting the Querent and also the future prospects for resolving the matter of the question.

Electional Astrology is used to pick a favorable time *in the fu-*

1897-1913. repr. 1968.), Book 4 (the House of Fortune and the House of the Spirit) and Book 6 (the other Houses—Father, Mother, Brothers, etc.).

[1]Al-Bîrûnî, *The Book of Instruction in the Elements of the Art of Astrology* ed & trans. by R. Ramsay Wright (London: Luzac, 1934.), pp. 279-295, gives a long list from Albumasar (c.786-886), *The Great Introduction*, Book VIII, and remarks sardonically that the number of Lots increases daily.

[2]Robert Zoller, *The Lost Key to Prediction: the Arabic Parts in Astrology* (New York: Inner Traditions International, c.1980). This book mentions a hundred or more Parts and gives instructions for their use.

ture to initiate some action, so that the action will have the best chance of producing the desired result.

The Alexandrian astrologers lumped Horary Astrology and Elections together under the Greek name *Catarchic* Astrology, which was the study and analysis of *beginnings* or *initiatives*. But medieval and modern astrologers have conveniently separated this joint category into Horary Astrology and Elections.

Astrologers have usually been consulted by clients wishing to learn some facts about the prospects for a particular region, a person, a situation, a question, or a proposed action. That is, *What are the present conditions?* and *What will happen next?*

For example, from an inspection of the *Caelum* at a particular moment in time, an astrologer can give a forecast of Mundane events likely to affect a particular region of the earth and the people, plants, and animals that live there. These can be prepared some time in advance, thus giving people an opportunity to prepare for the forecasted events.

From the Natal condition of the *Caelum*, the astrologer can describe the thinking and behavior of an individual, which may be useful for personal analysis or for the analysis of another person in whom the Client is interested. And the astrologer can also determine the probable success or lack of success that the Native will encounter in various areas of life in the future.

From a Horary chart, showing the condition of the *Caelum* at the moment a question is asked, the astrologer can forecast the attendant circumstances relating to that question and the probable outcome of the matter about which the Querent asked—in some cases, a simple "yes" or "no" answer.

A Horary chart can also be set for a past time when some event took place, in order to determine the attendant circumstances and the possible future conditions resulting from that event. This is technically called "a Horary chart Set for a Time Certain." One of these is called "the Decumbiture of the Sick." It is a chart set for the moment when a person first took to bed due to illness. It could be used to analyze the nature of his illness and to determine the prognosis for that person. It was much used in the 15th and 16th centuries when it was taught in the major universities and when medical doctors used it to diagnose a patient's malady, decide what treatment to give him, and make a prognosis as to whether or not he would recover.

Finally, if the Client wants to choose a propitious time to take some action in the future, the astrologer can make an Election, which means that he can select a time in the future when the *Caelum* will be most nearly favorable for whatever it is the client is proposing to do. This is also obviously a Horary chart Set for a Time Certain in the future.

In fact, it can be seen that all four branches of astrology are merely different versions of Horary astrology. A Mundane chart asks the question, "*What will happen next* at this location?" A Natal horoscope asks the question, "What is the personal character of this person?" and "*What will happen next* to him?" The Horary chart is set for an actual or an implied question, "*What will happen next* in this matter?" And an Election serves to answer the future question, "If I do it at that time, *What will happen next?*"

Thus we see that an astrologer could serve as a useful consultant for himself or for another person. In ancient Babylon, the king could consult an astrologer to interpret the *Caelum* to see what might affect his kingdom or himself personally. Initially, this amounted to the practice of Mundane astrology. But later it occurred to the Babylonian astrologers that the king's natal planetary positions could also be taken into consideration, since it was generally true that if conditions were favorable for the king, they were likely to be favorable for his subjects, and vice versa.

The Babylonian astrologers had compiled a large number of omens relating celestial conditions to mundane events, and they also paid particular attention to eclipses of the Sun and the Moon, which naturally aroused awe and wonder at the unusual event occurring in the sky. Their Natal astrology consisted of noting the sign position of the planets and sometimes whether they were close to a bright fixed star.

This service for the king naturally aroused envy in the hearts of other royal personages and noblemen of high rank. And then it must have spread to the upper middle class and eventually to the general population. Of the two dozen or more clay tablets that have been found to contain Babylonian horoscopes, two of the clients for whom the horoscopes were cast had Greek names, which shows that the art had already spread in the third century B.C. from the royal court to common citizens, and even to resident foreigners.[1]

And once the Babylonian astronomers and their Greek successors had learned how to predict mathematically where the Sun, the Moon, and the planets would be at any future time, the idea naturally occurred that if the motions of the celestial bodies were predictable, and if they had an influence on the people, plants, and animals of the earth, then the future of those living things should also be equally predictable. This idea was in harmony with the Stoic Philosophy developed by the Greeks.

As mentioned above, in the second century B.C. some astrologers in Alexandria, Egypt, combined Babylonian astrology, the Egyptian decans, and Greek religious and philosophical concepts to produce Horoscopic Astrology, which showed both the *state* of the *Caelum* and its *orientation* at a particular place and time.[2] This evidently became popular with the general public almost immediately. The Roman dictator Sulla (138-78) had his chart cast and received a favorable and useful reading, which to some extent influenced his actions during the rest of his life.

---

[1]See the recent publication by Francesca Rochberg, *Babylonian Horoscopes* (Philadelphia: American Philosophical Society, 1998). Text 10 records the birth of Aristocrates in 235 B.C., and text 12 records the birth of Nicanor in 230 B.C. The earliest of the 28 horoscopes discussed by Rochberg is dated to 410 B.C. and the latest to 69 B.C. See also Ulla Koch-Westenholz, *Mesopotamian Astrology* (Copenhagen: The Carsten Niebuhr Institute of Near Eastern Studies, 1995).

[2]The history of astrology from the Babylonian era down to modern times is set forth in detail by James Herschel Holden, *A History of Horoscopic Astrology* (Tempe, Arizona: American Federation of Astrologers, 1996), which contains many references to original sources and to historical works by other writers.

In September of 63 B.C., the Roman senator and astrologer Publius Nigidius Figulus (c.98-44 B.C.) announced to a startled fellow-senator Gaius Octavius that his newborn son Octavius would become the ruler of the world.[1] And as a young man of 18, the future Emperor Augustus (63-15 A.D.) received a very favorable reading of his natal horoscope from the astrologer Theogenes, and he evidently retained an interest in astrology for the rest of his life, even issuing a silver coin with his Moon sign Capricorn on it.[2]

Augustus's stepson the emperor Tiberius (42 B.C.-37 A.D.) employed astrologers as consultants, in particular his friend Tiberius Claudius Thrasyllus (d. 36), whose son Tiberius Claudius Balbillus (d.c. 82) became a life-long friend of the emperor Claudius (10 B.C.-54 A.D.). Balbillus also provided astrological counsel to the family of the emperor Nero (37-68) and to his successors down to the emperor Vespasian (9-79), who honored Balbillus by permitting the citizens of Ephesus to establish annual games in his honor.

In the early years of the first century A.D., astrologers began to notice that some aristocratic clients had configurations in their horoscopes that indicated prominence in public life. Then, carried away by enthusiasm, either the astrologers or the clients themselves dubbed the best of these "imperial horoscopes," by which they meant that those persons might become emperor. But for obvious reasons this seldom occurred, and the ruling emperor or members of his family often viewed persons with "imperial horoscopes" as a threat. The historians record that some of them were put to death by the ruling emperor or members of his circle.

The astrologer who adopted the impressive name Ptolemy Seleucus served as adviser to Nero's family and to the emperor Otho (32-69).[3] However, according to the historians, his advice

[1]Suetonius, *Lives of the Twelve Caesars*, ed. and tr. by Joseph Gavorse (New York: Modern Library, 1931), "The Deified Augustus," p. 109.

[2]Suetonius, "The Deified Augustus," p. 111.

[3]See the accounts of the historians cited by Frederick Cramer, *Astrology in Roman Law and Politics* (Philadelphia: The American Philosophical Society, 1954).

may not have been entirely helpful. The poet Juvenal (c.60-140) writing a half-century later spoke disparagingly of him.[1]

When Aulus Vitellius was born (probably on 7 September 15), "His parents were so aghast at his horoscope as announced by the astrologers, that his father tried his utmost while he lived, to prevent the assignment of any province to his son, and when he was sent to the legions and later hailed as Emperor, his mother immediately mourned him as lost."[2] She was right to do so. Eleven months later, in December 69, he was dragged from the palace in Rome, tortured, and murdered by an angry mob.

His successor, Vespasian, received some astrological encouragement from Balbillus, and he also seems to have acquired some personal understanding of astrology.[3] Suetonius says that "All agree that he had so much faith in his own horoscope and those of his family, that even after constant conspiracies were made against him he had the assurance to say to the Senate that either his sons would succeed him or he would have no successor."[4] He once scoffed at his younger son Domitian's reluctance to eat a dish of mushrooms by telling him that he was destined to die by the sword rather than by poison. And Domitian (51-96) was later the recipient of an unfavorable astrological forecast of his death at a certain

---

[1]Juvenal, *Satires* (in *Juvenal and Persius*, tr. by G. G. Ramsay, Loeb Classical Library, (Cambridge, Mass.: Harvard University Press, 1918; rev. ed. 1940.), vi. 557-559. A few lines down (vi. 566) the poet mentions the name of Tanaquil, wife of King Tarquinius Priscus (reigned 616-578 B.C.) as the type of a lady well-informed about astrology. He probably had in mind that the historian Livy had said of her (*History*, i. 34) that she was *perita caelestium prodigiorum* 'expert in celestial prodigies', but her date is too early for the prodigies to be astrological in nature. Probably the reference was to her interpretation of atmospheric phenomena and the flight of birds. In the same part of the satire, Juvenal also provided a testimonial to the pervasiveness of astrology among the Romans of his day (c.115) when he cautions the reader to beware of a woman carrying a well-rubbed ephemeris with her and who governs her everyday activities by astrological indications.

[2]Suetonius, "Vitellius," 308.

[3]Cramer, 135-139.

[4]Suetonius, "The Deified Vespasian," 336.

time (18 September 96 at 11 A.M.),[1] which actually occurred precisely as forecasted, when he was murdered by two of his familiars.[2]

The emperor Hadrian was born at Italica (near Santiponce), Spain, at about 6:40 AM LAT on 24 Jan 76. His great-uncle the senator Aelius Hadrianus, who was reputed to be a master of astrology, had predicted that Hadrian would become emperor. Later, when he was serving as tribune of the Fifth Legion in Lower Moesia (now, northern Bulgaria), he received the same prediction from a local astrologer. The historians tell us that he was of a scholarly turn of mind and that he became so expert in the art of astrology "that on the Kalends of January he would actually write down all that might happen to him in the whole ensuing year, and in the year in which he died, indeed, he wrote down everything that he was going to do, down to the very hour of his death."[3]

It is recorded that the emperor Commodus and his twin brother Antoninus were born on 31 August 161 and that for the latter "the astrologers had cast a horoscope as favorable as that of Commodus," but that he "lived to be only four years old."[4] But the

[1]Suetonius, "Domitian,." 335-336. "He had long since had a premonition of the last year and day of his life, and even of the very hour and manner of his death. In his youth astrologers had predicted all this to him, and his father once even openly ridiculed him at dinner for refusing mushrooms, saying that he showed himself unaware of his destiny in not rather fearing the sword."

[2]Sceptics have argued that this was a "self-fulfilling prophecy." But the fact that the prediction was publicly known does not alter the fact that it was made some years in advance and came true on the forecasted date despite Domitian's efforts to avoid it.

[3]*Scriptores Historiae Augustae*, tr. by David Magie 3 vols. Loeb Classical Library (Cambridge, Mass. and London, 1921), "Hadrian," xvi. 7. Fortunately, Hadrian's horoscope and a commentary by Antigonous of Nicaea is preserved in Hephaestio of Thebes, *Apotelesmatica* ed. by David Pingree, 2 vols. (Leipzig: B. G. Teubner, 1973-74), vol. 1, ii. 18. See my translation of the Greek text and the horoscope diagrams in Holden, 58-61; also, Cramer, 162-178, where the Greek text and horoscope diagram are reproduced in facsimile from the MSS; and Neugebauer, O. and Van Hoesen, 90-91. Modern astrologers would note with interest that Hadrian had Uranus in the same sign (Aquarius) with his Sun.

[4]*Scriptores Historiae Augustae*, "Commodus Antoninus," 265.

astrologers' forecasts were wrong on both counts, since Antoninus died early and Commodus was eventually murdered at age 31.

The emperor Septimius Severus (145-211) consulted astrologers in his youth, received a favorable forecast (he had the Sun, Moon, and Venus in their exaltation signs and Jupiter in his own sign Sagittarius), and also learned the art himself. He later employed astrology in an interesting way by having a search made for a woman he had heard of whose chart indicated that she would have an imperial husband. Having located that person, Julia Domna (d. 217), in Syria, he married her in 187, and six years later he became Emperor. He is said to have had his horoscope painted on the ceiling of one of his private rooms in the palace, so that he could consult it whenever he wanted to.[1]

There were undoubtedly some other astrologers who were consultants to later Roman Emperors, although the skimpy histories of the later Empire do not give details.

Unlike most other nations, the Romans always celebrated the ruling Emperor's birthday, and probably the Emperor's age was generally known, so astrologers interested in making Mundane predictions at least knew the current Emperor's birthdate.[2] In some cases his birth time was also known, so it was possible to speculate on the date of his demise. Naturally, such speculations came to the reigning emperors' attention, and most of them were not amused. This gave rise to several edicts against the practice of astrology, but they were seldom enforced—astrology was too popular with the public. The fourth century Roman astrologer Julius Firmicus

---

[1]Astrology in the Roman World down to 200 A.D. is discussed in detail by Cramer. He is said to have collected notes for a similar volume on the later Empire, but he died before these were converted into a book and published.

[2]A good example of this is the horoscope of the emperor Nero, which is given as an example by Vettius Valens, *Anthologiae* ed. David Pingree (Leipzig: B. G. Teubner, 1986), v. 7. 20-35 "The Aphetic Place", of course without any mention of the emperor's name. As I pointed out in my book, Holden, 53-55 (where a translation of the passage is given), neither Neugebauer & Van Hoesen, *Greek Horoscopes*, nor Pingree in his edition of the *Anthologiae* recognized that this was the horoscope of Nero!

Maternus warned the readers of his astrological manual not to attempt to read the emperor's horoscope, for, as he says, the emperor is not subject to the decrees of the stars.[1]

We should mention that both the astrologers and the public believed that the natal horoscope indicated the native's fate in life—not just the future conditions but also the pre-existing conditions and those in existence at the time of birth. Actually, this is only partly true. Claudius Ptolemy, *Tetrabiblos* i. 2, clearly states that genetics and social circumstances must be taken into account before making a forecast for a particular individual. He adds that those who are ignorant of their destiny will certainly experience it, but those who know it and have the courage to take steps to avoid it may succeed at least in part.[2]

But Ptolemy's book did not become available to astrologers until the beginning of the fourth century, and even when they had it, that dictum was largely ignored.[3] The Christian writer St. Augustine (354-430), who had studied astrology in his younger years, says that he gave up his belief in astrology when he heard that the son of a wealthy nobleman and the son of a slave had been born at the same hour.[4] But those who had read Ptolemy's remarks would not have had a problem with that.

[1]Julius Firmicus Maternus, ii. 30.5 (p. 86 in the Teubner edition). See also the English translation of the whole chapter in Holden 66-70.

[2]A convenient modern edition is *Ptolemy/ Tetrabiblos* ed. and tr. by F.E. Robbins in the Loeb Classical Library (London and Cambridge, Mass., 1940).

[3]In my unpublished paper on Claudius Ptolemy's place in society, I point out the significant fact that his younger contemporary Vettius Valens (120-c.175) lived in Alexandria during the latter part of Ptolemy's life and had never heard of him or his books! The first astrological writing to mention Ptolemy seems to have been Porphyry. *Introduction to the Tetrabiblos* ed. by Emilie Boer and Stephen Weinstock in *Catalogus Codicum Astrologorum Graecorum* (Brussels: Belgian Academy, 1940), 5, 4. It was written a few years before 300 A.D. And even Julius Firmicus Maternus, writing in the mid-fourth century, did not have Ptolemy's books.

[4]See Lynn Thorndike, *History of Magic and Experimental Science* 8 vols. (New York: Columbia University Press, 1923-1958.), vol. I, Chapter XXII "Augustine on Magic and Astrology."

At this point we should describe the techniques employed by the early astrologers to forecast future events.

The Babylonians seem to have mainly relied upon their extensive collection of omens, a careful consideration of eclipses, and later some rudimentary significations of the planets in the various signs of the zodiac. They had available arithmetical schemes for calculating the positions of the Sun, Moon, and planets, step by step; and in some cities there may have been archives or libraries that contained ephemerides on clay tablets.

With the invention of the Horoscopic Astrology chart in the second century B.C. and the several classifications of the signs of the zodiac mentioned above, the Alexandrian astrologers had available a much more elaborate set of criteria by which they could judge the signification of the *Caelum* at any particular time and place. They were thus able to analyze the Natal positions in much more detail.

According to theory, the horoscope chart represented both the present and future of the person or thing. But to estimate when something indicated in the chart might manifest itself at any subsequent time, they seem to have employed two tools: (1) a sort of combination numerological and astrological scheme of successive time periods; and (2) what are now called primary zodiacal directions of the Ascendant, which amounted to a rotation of the original chart in the clockwise direction by one degree per year. The first of these tools provided a definite succession of planetary rulers for each future time period down to the hour; the second provided a progression of the ascending degree through the zodiac which altered its rulerships and brought it into different "terms," signs, or into aspect with the planets. Both of these sets of indications were judged with respect to the nature and quality of the original chart.

To supply the astronomical data needed for these charts, a set of tables called the *Eternal Tables* had been prepared at Alexandria.[1]

[1]I believe these tables are first mentioned in Neugebauer and Van Hoesen, "Original Documents, " No. 81 (London Papyrus 130), where

These were used for nearly 500 years until Ptolemy's more accurate tables began to become available in the fourth century. The initial degree of precision of the *Eternal Tables* was probably within 2 or 3 degrees of the truth, but with the passage of time the errors grew greater. Fragments of ephemerides on papyrus have been found, which indicates that they may have been available in Alexandria or elsewhere on a limited basis. But most astrologers probably used the *Eternal Tables* to calculate horoscopes. Those tables referred longitudes to the meridian of Alexandria and to the fixed Alexandrian zodiac.

For persons born at points distant from Alexandria, the careful astrologer could consult some geographical work that stated that Rome was 1 hour west of Alexandria, Spain was 2 hours west, Babylon was 1 hour east, etc. The Ascendant was determined from the hour of the day or night, the longitude of the Sun, and the *clime* of the place, initially using arithmetically determined rising times of the individual signs, and finally using the much more accurate rising times determined trigonometrically by Ptolemy in his *Handy Tables*.[1] Those tables also offered more precise positions of the planets, and they were referred nominally to the tropical zodiac.[2] House positions in the horoscope were generally established according to the Sign-House system, in which the ascending sign was the first house, the next sign the 2nd house, the one after that the 3rd house, etc. Four other methods of determining the houses were invented, but they never came into general use during the Classical Period.[3]

---

the planetary positions for a horoscope set for 31 March 81 A.D. are said to have been calculated *dia kanonôn aioniôn* 'by eternal tables'.

[1]These were edited and translated into French by the Abbé Nicolas Halma, *Tables manuelles des mouvemens des astres*, 3 vols. (Paris, 1822-25).

[2]Unfortunately, Ptolemy had adopted 36″ per year for the value of the annual precession instead of the true value of nearly 50″ per year. And since the true epoch of his longitudes was about 130 B.C., the "tropical" longitudes yielded by his tables were about 2° in error by the end of the fourth century.

[3]These techniques are discussed by Holden, 'First Period: Babylonian

By using the two techniques mentioned above, the astrologer could prepare for a client a forecast of future conditions and events. Or, if he was judging a Horary chart to give an answer to a Client's question, he could employ some similar techniques. Available handbooks explained the rules for judging the different types of charts.

Obviously, if the client was uncertain about his birthdate, or if he knew the date but did not know the time, it was not possible to erect a natal horoscope for him. In such a case, the astrologer might forgo the natal chart and derive what information he could from the planetary positions at say noon on the day of birth. Or, he could consider the personality of the client and guess what ascendant he might have. Then he could make a *speculative horoscope* for him. This is occasionally still done—sometimes with success but more often not.

If the astrologer wanted to make a Mundane forecast, he usually considered a chart set for the time when a New Moon or a Full Moon most closely preceded the entrance of the Sun into the sign Aries each year. And he also considered the planetary positions associated with any eclipse that might occur during that year. Definite rules enabled him to judge the weather, natural events, agricultural conditions, and social circumstances during that year.[1]

These were the usual astrological tools employed by astrologers from the second century B.C. down to the end of the Classical Period. Ptolemy had explained the use of what we would now call "Placidian primary directions *in zodiaco*" in Book 3 of his *Tetrabiblos*, but the book only began to become available in the fourth century, and that was too late for that technique to come into common use in the Classical Period.

With the collapse of the Roman Empire in the West in the fifth century and the general decline of literacy, astrology virtually died out in Western Europe. It was after all a Greek science, since most

---

Astrology' and 'Second Period: Classical Greek Astrology'.

[1]The procedures are explained by Ptolemy, Book 2.

of the textbooks were in Greek. It also declined in Byzantine territory until the 10th century, when a revival began that lasted until 1453.

In Muslim territory there were a number of rulers who employed astrologers as consultants. Nawbakht the Persian (c.679-777) was court astrologer to the Caliph al-Mansûr (d. 775). He was selected to head a group of astrologers, including the young Jewish astrologer Mâshâ>allâh (c.740-c.815) to select a date and time for founding the city of Baghdad in the summer of 762 A.D.[1]

Another early astrologer was Theophilus of Edessa (c.695-785) who was court astrologer to the Caliph al-Mahdî (d.785). Following him, Nawbakht's son Abû Sahl ibn Nawbakht (d.786) succeeded him as court astrologer, and his son Abû Sahl al-Fadl ibn Nawbakht (c.735-815) was court astrologer to the famous Caliph Hârûn al-Rashîd (reigned 786-809). Abû Sahl's descendants served as court astrologers to the Caliphs al-Ma>mûn (reigned 813-833), al-Wâthiq (reigned 842-847), and al-Mutawakil (reigned 847-861). Five generations of court astrologers from the same family!

Astrology flourished in the Muslim territories during the eighth and ninth centuries, but thereafter it declined steadily due to increasing religious intolerance, although it hung on longer in North Africa and southern Spain.

Later, in Western Europe, astrology began a strong revival in the 12th century with the translation of Arabic astrological trea-

---

[1]See my paper "The Foundation Chart of Baghdad" in the American Federation of Astrologers periodical *Today's Astrologer*, Vol. 65, No. 3 (March 2, 2003). The chart given there is based on the chart preserved in Al-Bîrûnî, *The Chronology of Ancient Kingdoms* tr. by Eduard Sachau (London, 1879), p. 263. The adjacent text puts the foundation on the 23rd of Tammûz 1074 Alexander, which is equivalent to 23 July 763, and says that the constellations of the heavens "are indicated in the following figure." But either the date is wrong or the figure is wrong, since the figure is set for 31 July 762. I am inclined to think that the figure is more likely to have been transmitted correctly than the date.

tises into Latin and the rise of literacy. In the next century, we hear of astrologers who were consultants to rulers, ecclesiastics, and the merchant class. The famous Italian astrologer Guido Bonatti (c.1210-c.1295) was one of these. In his younger years he was attached to the court of the Roman Emperor Frederick II (1194-1250). In September 1261 he judged a question about an impending battle for Count Guido Novello (d. 1266), the Podestà of Florence, and forecast that he would win the battle, which he did. Bonatti also composed a massive treatise on astrology entitled *A Book of Introduction to the Judgments of the Stars*, which was printed at Augsburg by Erhard Ratdolt in 1491.[1]

In the Eastern Empire, John Abramius was court astrologer to the Byzantine Emperor Andronicus IV (reigned 1376-1379).

Greek horoscopic astrology had passed to the Persians, the Hindus, and later to the Arabians beginning in the second century A.D. Each of these ethnic groups modified the original Alexandrian rules and procedures somewhat to suit their own notions. Persian astrology was ultimately absorbed by the Muslims. Hindu astrology remained mainly independent and gradually deviated to a considerable degree from the original Greek astrology, although it retained the basic framework, most notably clinging to a fixed zodiac.

The Muslims continued to use most of the procedures mentioned in the Greek astrological manuals, but they introduced two new features that have since become standard tools—the Aries Ingress for Mundane astrology and the Solar Return for Natal astrology. Having (rightly or wrongly) more faith in the accuracy of their astronomical tables, they discarded the consideration of the New or Full Moon preceding the Aries Ingress and instead erected a chart for the ephemeris time of the Ingress itself. And they introduced the idea of erecting a chart annually for the moment when the Sun returned to the exact degree and minute that it occupied at

[1]Guido Bonatti, *Guido bonatus de forlivio: Decem continens tractaus astronomie* (Augsburg: Erhard Ratdolt, 1491). In the MSS the book is called *Liber introductorius ad iudicia stellarun*.

a person's birth. The resulting chart, originally called a "revolution" (now termed "a solar return"), was judged along with the Natal chart to assess the prospects for the individual during the ensuing year.

They also continued the use of zodiacal primary directions. And they utilized spherical trigonometry more assiduously than their predecessors had in the Classical Period. In the case of house division, they utilized the Sign-House system of the Alexandrian astrologers and also the so-called Alchabitius system that actually goes back to the close of the Classical Period.[1]

However, Muslim astrologers were hampered by the lack of accurate birth data. There were no public registers of births, and there was evidently little or no precise biographical information in the libraries from which one could discover the birth data of the king or other important persons. (It is significant that no collections of nativities have come down to us from the astrologers writing in Arabic.) If the client was unable to provide birth data for himself or those he was interested in, the astrologer had to fall back upon Horary charts and perhaps Mundane charts. From reading the surviving books written by Muslim astrologers, it would appear that Horary astrology was the principal tool they used for prediction.[2]

The situation just described was also true in the West and lasted there until the 15th century, when the European literacy rate had considerably improved and records had become more plentiful. Also, the invention of printing made more books available and greatly reduced their prices. Many of the Latin translations of Arabic astrology books that had been made in the 12th century were printed for the first time after about 1480. Julius Firmicus Maternus's fourth century text the *Mathesis* was printed twice in

---

[1]See my paper "Ancient House Division" in the American Federation of Astrologers *Journal of Research*, Vol. 1 (1982): 19-28, and the Italian version "Antichi sistemi di domificazione" in *Linguaggio Astrale* 97 (1994), (Torino: CIDA).

[2]See Holden, "The Third Period: Medieval Astrology."

the 1490s, and some of the medieval astrological treatises by writers such as Guido Bonatti began to appear. But the Greek text of Ptolemy's *Tetrabiblos* was not published until 1535, although one of the medieval Latin translations from the Arabic version had been available in print since 1484. And unfortunately some astrologers both then and later mistakenly assumed that Ptolemy was the inventor of astrology and that his *Tetrabiblos* was the prime authority.

The German astronomer and astrologer Johann Müller (1436-1476), better known as Regiomontanus, misunderstood Ptolemy's instructions for calculating primary directions and devised an entirely different method, which he also applied to the calculation of house cusps. He published an ephemeris at Nürnberg in 1474, perhaps the first printed ephemeris, and he or his assistant Martin Bylica (1433-1493) also prepared an extensive set of tables for making the necessary calculations for horoscopes and primary directions.

These tables, finally published at Augsburg,[1] in 1490, were the first printed set of house and auxiliary tables. Almost immediately the Regiomontanan house system replaced the Alchabitius system and, became the standard used by astrologers (with a few exceptions) for the next two centuries. Printed maps also became available, which provided astrologers with reasonably correct geographic latitudes and longitudes; thereby improving the accuracy of their charts. Clocks began to become available for those who could afford them, and sundials or taking note of church bells provided the time to the less fortunate.

More ephemerides were published in succeeding years, thus relieving those astrologers who could afford them of the necessity of using tables to calculate solar, lunar, and planetary positions. Unfortunately, the inherent error in the tabular longitude of the Sun caused the time of all of their Aries Ingress charts to be in error by

[1] *Tabulae directionum profectionumque...in nativitatibus multum utiles.* [Tables of Directions and Profections...very Useful in Nativities] (Augsburg : Erhard Ratdolt, 1490.); often reprinted in later years.

several hours, thus vitiating the analysis based upon house positions, and injecting errors into their mundane predictions. And the longitude of Mercury could sometimes be in error by as much as 8 or 10 degrees.

Astrology blossomed in the Renaissance. There were chairs of astrology in the major universities. It was part of the curriculum for students of medicine. The Italian astrologer Luca Gaurico (1476-1558) twice predicted to Cardinal Alessandro Farnese (1468-1549) that he would become Pope. When this occurred in 1534, Farnese, as Pope Paul III, summoned Gaurico to Rome, made him his table companion, knighted him, and further rewarded him in 1539 by creating him Bishop of Giffoni. Gaurico was later court astrologer to the Queen of France, Catherine de' Medici (1519-1589).[1]

Jerome Cardan (1501-1576) was a famous physician, mathematician, astrologer, and miscellaneous writer. He was consulted by many important people, including John Hamilton (c.1511-1571), the Archbishop of St. Andrews in Scotland, who was gravely ill. Cardan diagnosed him as having what we today would call an allergy to feathers, and he prescribed effective remedies. The Archbishop begged him to stay as his personal physician, but Cardan chose not to do so.

On his way home, Cardan was introduced to the young King Edward VI (1537-1553) of England. Cardan examined him physically and also cast his horoscope. It looked promising, he told the royal officials. Unfortunately, the King later contracted pneumonia and died in July 1553. Cardan was so chagrined at his failure to see that in the King's horoscope that he later published a short tract discussing the chart and speculating on the possible reasons for his error.[2] Modern astrologers can see why he made his mistake. He

[1]Thorndike, vol. V, Chapter XIII "The Court of Paul III."

[2]See Jerome Cardan, *Opera omnia*, 10 vols. (New York and London: Johnson Reprint Corporation, 1967; a facsimile reprint of the Lyons ed. of 1663), vol. 5, 503 ff., "*Liber duodecim geniturarum*" [The Book of Twelve Nativities]. The first nativity is that of King Edward VI.

had "rectified" the King's stated birth time of 2 A.M. back to about 1:30 A.M. using a procedure advocated by Claudius Ptolemy. That gave the chart a 29 Leo ascendant instead of the 7 Virgo ascendant, which would have revealed the King's danger from serious illness.

The German Reformation scholar Philip Melanchthon (1497-1560), a close friend and associate of Martin Luther (1483-1546), was the first to edit the text and translate Ptolemy's *Tetrabiblos* directly from Greek into Latin (Basel, 1553). He was especially interested in Luther's horoscope (the exact date of which had been in doubt at first).[1] Unlike many other Protestant theologians, Melanchthon saw no conflict between astrology and religion.

In England, Dr. John Dee (1527-1608) was court astrologer to Queen Elizabeth I (1533-1603). He signed his dispatches to her with the monogram "007," in which the horizontal part of the 7 was extended over the 00 in the manner of a lorgnette to indicate that he was "the eyes" of the Queen. (This is the origin of Ian Fleming's designation for his fictional secret agent James Bond.)

Dee's son Arthur Dee (1579-1651) was recommended to the Russian court by King Charles I (1600-1649) of England. Arthur served as physician and court astrologer to Tsar Michael Feodorovich Romanov (1596-1645) for a number of years before returning to England.

One of the modern biographies of Tsar Peter the Great (1672-1725) records that the astrologers said that "they had seen his star rise in the immediate neighborhood of the planet Mars." (And he is said in another biography to have been born at 5 AM, which would put Mars in the Midheaven.) This report would seem to indicate that there were astrologers at the court of Tsar Alexis I Mikhailovich (1629-1676), Peter's father.

The famous astronomer Tycho Brahe (1546-1601) was court

[1]The problem experienced by the Renaissance astrologers in obtaining correct birthdata for Martin Luther is discussed at length by Will-Erich Peuckert, *Astrologie* (Stuttgart: W. Kohlhammer, 1960), in the chapter "Die Stunde der Geburt" [The Hour of Birth], 114 ff.

astrologer to the Danish King Frederick II (1534-1588). He took special interest in the "new star" that was first seen on 6 November 1572. Having considered its astrological implications, he wrote a book[1] in which he predicted that a great warrior would be born in the North not before 1592, that he would conquer Germany, and would then disappear in 1632. This was later borne out in the person of the Swedish warrior king Gustavus II Adolphus (1594-1632).

Tycho's sometime assistant and successor, the renowned astronomer John Kepler (1571-1630) served as court astrologer to the emperor Rudolph II (1552-1612), whom he honored by naming his astronomical tables the *Rudolphine Tables*, and further honored by using the Emperor's horoscope as an example in the tables. He was also consulted on at least two occasions by the famous general Count Wallenstein (1583-1634), who had a personal astrologer Giovanni Battista Seni (1600-1656) of Padua, who traveled with him and gave him astrological counsel in the years 1629-1634.

Astrology was greatly benefited by Kepler's discovery of elliptic orbits and by the publication of his *Rudolphine Tables*, which for the first time in history gave astrologers fairly accurate planetary positions for their horoscopes, although the Sun had a cyclic error of about 7 minutes of arc, and the Moon's error could still amount to half a degree at times.

In Italy, Galileo Galilei (1564-1642), justly famous for his discoveries in physics, and in astronomy with the newly invented telescope, also drew his own horoscope and the horoscopes of his children and served as court astrologer to Cosimo II (1590-1621), the Grand Duke of Tuscany.[2]

In France, the physician and astrologer Jean Baptiste Morin

---

[1]Tycho Brahe, *De nova et nullius memoria prius visa stella*... [On the New and in No One's Memory Hitherto Seen Star...] (Copenhagen: Laurentius, 1573).

[2]See the papers "Gli 'Astrologia nonnulla di Galileo' I & II" by Serena Foglia and Grazia Mirti in *Linguaggio Astrale* 88 (Turin: CIDA, 1992).

(1583-1656) was well-known to the courtiers of King Louis XIII (1601-1643). He was appointed Regius Professor of Mathematics at the Collège de France, and was summoned to be present at the birth of the future King Louis XIV (1638-1715), in order to prepare an accurate horoscope for him. Morin was frequently consulted by Cardinal Richelieu (1585-1642), who disliked him personally, but had faith in his astrological skill.

Morin spent the last 30 years of his life composing a massive astrological treatise *Astrologia Gallica*, which unfortunately he never lived to see in print. It was published in 1661 at the expense of a grateful former client, Queen Marie Louise of Poland (1611-1667).[1] In his book, Morin set forth explicit methods of judgment that differed considerably from those in common use in the 17th century (his methods were in some respects similar to those used in Horary Astrology). Morin's methods never came into use in that century, since astrology was on the wane in France, but they were revived at the beginning of the 20th century and are slowly becoming known to astrologers today.[2]

During the Cromwell era in England, from the outbreak of the Civil War in 1642 until the Restoration in 1660, the English astrologers chose sides, and in their annual almanacs and other publications they either supported the royalists or, more commonly, the Parliamentarians. Chief among these was the famous William Lilly (1602-1681), who favored the Parliamentarians, while his younger rival John Gadbury (1627-1704) sided with the royalists. The leading astrologer and almanac maker at the end of the 17th century and the beginning of the 18th century was John Partridge (1644-1715).

The principal astrological tools used by the astrologers of the early modern period were Natal horoscopes with their primary di-

---

[1]Morin, Jean Baptiste, *Astrologia Gallica* (The Hague: A. Vlacq, 1661). folio.

[2]Holden, 164-168. In addition to the translations cited there, my translations of Morin, AG Books 23-25, will be published by the AFA in 2004-5, which has also published a translation of Book 18 by Pepita Sanchez Llacer and Anthony Louis LaBruzza.

rections (produced by rotating the chart at the rate of 1 degree per year), solar returns (charts set for the moment each year when the Sun returns to its natal longitude), lunar returns, and transits, and Horary charts for answering questions. In the field of Mundane Astrology, they used Aries Ingresses, charts set for lunations (especially eclipses), and charts related to conjunctions of the outer planets. Many astrological almanacs were printed containing predictions of the weather and various kinds of mundane events.

We must also mention the Italian professor of mathematics Placidus de Titis (1603-1668), generally known simply as Placidus, who served as an astrological consultant to the Archduke Leopold William of Austria (1614-1662). Unlike Regiomontanus, Placidus understood Ptolemy's method of primary directions. He wrote a treatise explaining both zodiacal and mundane primaries (now called Placidian primaries) and also devised a new system of house division, which has since been called the Placidus system.[1] It was adopted by the English astrologer John Partridge in the late 17th century, and by most English astrologers by 1800. Its use spread from England to Europe in the 20th century, where it generally replaced the Regiomontanus system (except in Austria).

Placidus also introduced what he called "secondary progressions" as an auxiliary predictive tool. But as augmented two and a half centuries later by the English astrologer Alan Leo (1860-1917) under the name "The Progressed Horoscope," secondary progressions and a version of zodiacal primaries have become the principal predictive tools used today in conjunction with transits.[2] Primary directions and solar and lunar returns lost favor,

---

[1]This system was first proposed by Giovanni Antonio Magini (1555-1617), *Primum Mobile* (Venice, 1604), but it was Placidus who popularized it a half century later. After some earlier tracts, his methods were set forth fully with examples in his *Quaestionum physicomathematicarum libri tres*... (Milan, 1650) and his *Tabulae Primi Mobilis*... (Padua, 1657). The latter book was translated into English by John Cooper as *Primum Mobile*... (London: Davis and Dickson, 1814); reprinted in facsimile with an Introduction by Michael Baigent (London: Calverts Press, 1983).

[2]See my paper in the AFA *Journal of Research*, Vol. 11 (Winter 1999),

since they were harder to calculate, but have recently recovered somewhat due to the advent of computers, which make their calculation effortless.[1]

Interest in astrology largely died in Continental Europe towards the end of the 17th century, not to be revived until the 1890's. In England it also declined but was kept alive by the continuing public interest in annual prophetic almanacs, until natal astrology came back to life around 1785. Thereafter, it grew steadily in popularity, culminating with the books and magazines issued by Alan Leo and his publishing house Modern Astrology. Leo was a Theosophist, and since Theosophy favored astrology, that increased the size of the audience for his books, which were soon translated into several other languages, thus helping to spark a European revival of the art in the period 1890-1930. Since that time, astrology has become increasingly popular among Europeans and their descendants around the globe.[2]

During the 18th and 19th centuries, astronomers greatly improved the accuracy of the ephemerides that constitute the principal working tool for astrologers. In particular, they first succeeded in producing solar and lunar positions accurate to 1 minute of arc and eventually accurate to 1 second. This increased the precision of astrological prediction.

In recent times, the advent of the personal computer has made astronomical computation accurate and effortless. And the wealth of astronomical data available on the Internet has also become available to astrologers. One result of this is a recent upsurge of interest in Asteroids, which had been largely ignored by earlier astrologers due to the lack of ephemerides.

During the period 1700 to the present, I have not found any notice of official court astrologers in Europe, although there were a few individual aristocrats who had their horoscopes cast and

---

61-64 "The Progressed Horoscope: A Mixed Bag":

[1]Holden, "Fourth Period: Early Modern Astrology."

[2]Holden, "Fifth Period: Modern Astrology" and "Some Remarks on Astrology".

sought astrological advice. It is said that some of Hitler's lieutenants were interested in astrology, although in the last years of the Nazi era the German government became afraid of astrological predictions and ordered the astrologers to be detained and their magazines suppressed. But after World War II, astrology rapidly recovered its popularity in Germany and the rest of Europe.

The late American President Ronald Reagan (b. 1911) was the victim of an attempted assassination in 1981. Thereafter, his wife Nancy Reagan (b. 1921) employed the San Francisco astrologer Joan Quigley to supply a list of dangerous times for the President, and his schedule of personal appearances was arranged to avoid those times.

In Asia, astrology was always very popular in India and China under their native monarchs. Many of them had court astrologers, but their names must be sought in the histories of those countries. It should be mentioned, however, that astrologers are frequently consulted by all ranks of modern Indian society, and astrology is again becoming popular in the world's largest nation, China. However, with the upsurge of fundamentalism in Muslim countries, it has once again been subjected to religious criticism.

In Western society around the world the popularity of astrology has steadily risen. Newspaper and magazine "Daily Guides" are read by tens of millions; more people are studying astrology; many new books on astrology are published each year; the number of professional astrologers has grown; some colleges and universities are offering courses in astrology; and the number of clients consulting astrologers has shown a steady increase.

Still, it must be remembered that all of this is due to the age-old persistent question in people's minds: **What will happen next?**

References

Al-Bîrûnî. *The Book of Instruction in the Elements of the Art of Astrology* ed. and tr. by R. Ramsay Wright (London: Luzac, 1934.).

Al-Bîrûnî. *The Chronology of Ancient Kingdoms* tr. by Eduard Sachau (London, 1879).

Albumasar (Abû Ma<shar). *Introductorius maior . . . in astrologia . . .* [The Great Introduction to Astrology] tr. by Hermann of Carinthia (Augsburg: Erhard Ratdolt, 1489).

Bonatti, Guido. *Guido bonatus de forlivio: Decem continens tractatus astronomie*. (Augsburg: Erhard Ratdolt, 1491).

Brahe, Tycho. *De nova et nullius memoria prius visa stella* . . . [On the New and in No One's Memory Hitherto Seen Star . . .] (Copenhagen: Laurentius, 1573).

Cardan, Jerome. *Opera omnia*, 10 vols. (New York and London: Johnson Reprint Corporation, 1967; a facs. repr. of the Lyons ed. of 1663).

Firmicus Maternus, Julius. *Mathesis,* 2 vols. (Leipzig: B.G. Teubner, 1897-1913. repr. 1968.)

Cramer, Frederick. *Astrology in Roman Law and Politics* (Philadelphia: The American Philosophical Society, 1954).

Foglia, Serena and Grazia Mirti. *Linguaggio Astrale* 88 (1992).

Ginzel, F. K. *Handbuch der mathematischen und technischen Chronologie*, 3 vols. (Leipzig: J. C. Hinrichs, 1914.).

Halma, Abbé Nicolas. *Tables manuelles des mouvemens des astres*, 3 vols. (Paris, 1822-1825).

Heliodorus. *In Paulum Alexandrinum commentarium* [Commentary on Paul of Alexandria] ed. by Emilie Boer (Leipzig: B. G. Teubner, 1962).

Hephaestio of Thebes. *Apotelesmatica*, ed. by David Pingree 2 vols. (Leipzig: B. G. Teubner, 1971).

Holden, James Herschel. *A History of Horoscopic Astrology* (Tempe, Arizona: American Federation of Astrologers, 1996).

Journal of Research of the A.F.A. (Tempe, Arizona: A.F.A., Inc., 1982-

Juvenal. *Satires* trans. by G. G. Ramsay in *Juvenal and Persius*, Loeb Classical Library, 1918, rev. 1940.

Koch-Westenholz, Ulla. *Mesopotamian Astrology* (Copenhagen: The Carsten Niebuhr Institute of Near Eastern Studies, 1995).

Linguaggio Astrale (Turin: Centro Italiano di Astrologia, 1970-).

Magini, Giovanni Antonio. *Primum Mobile* (Venice, 1604).

Morin, Jean Baptiste. *Astrologia Gallica* (The Hague: A. Vlacq, 1661).

Neugebauer, O. *Astronomical Cuneiform Texts*, 3 vols. (London: Lund Humphries, 1955).

Neugebauer, O. and H. B. Van Hoesen, *Greek Horoscopes* (Philadelphia: American Philosophical Society, 1959), pp. 90-91.

Paul of Alexandria. *Eisagôgika* [Introduction], ed. by Emilie Boer (Leipzig: B. G. Teubner, 1958.).

Peuckert, Will-Erich. *Astrologie* (Stuttgart: W. Kohlhammer, 1960).

Placidus de Titis, Didacus. *Quaestionum physicomathematicarum libri tres . . .* (Milan, 1650)

Placidus de Titis, Didacus. *Tabulae Primi Mobilis...* (Padua, 1657).

Placidus de Titus, Didacus. *Primum Mobile* tr. into English by John Cooper (London: Davis and Dickson, 1814); repr. in facs. with an intro. by Michael Baigent (London: The Institute For the Study of Cycles in World Affairs, 1983).

Porphyry. *Introductio in Tetrabiblum Ptolemaei* ed. by Emilie Boer and Stephen Weinstock in *Catalogus Codicum Astrologorum Graecorum* 5.4 (Brussels: Belgian Academy, 1940).

Ptolemy. *Tetrabiblos* ed. and tr. into English by F. E. Robbins, Loeb Classical Library (London and Cambridge, Mass., 1940).

Regiomontanus, Johannes. *Tabulae directionum profectionumque . . . in nativitatibus multum utiles.* [Tables of Directions and Profections . . . very Useful in Nativities] (Augsburg: Erhard Ratdolt, 1490. 4to.

Rochberg, Francesca. *Babylonian Horoscopes* (Philadelphia: American Philosophical Society, 1998).

Scriptores Historiae Augustae, translated by David Magie 3 vols. (Loeb Classical Library, London and Cambridge, Mass., 1921).

Suetonius, *Lives of the Twelve Caesars* ed. and tr. by Joseph Gavorse (New York: Modern Library, 1931), p. 109.

Thorndike, Lynn. *History of Magic and Experimental Science* 8 vols. (New York: Columbia University Press, 1923-1958).

Vettius Valens. *Anthologiae* ed. D. Pingree (Leipzig: B. G. Teubner, 1986).

Zoller, Robert. *The Lost Key to Prediction: the Arabic Parts in Astrology* (New York: Inner Traditions International, c.1980).

THE HOROSCOPE OF THE EMPEROR TIBERIUS

By James Herschel Holden, M.A. FAFA

The second emperor of Rome, Tiberius, was the son of Tiberius Claudius Nero (d. 38 B.C.) and his wife Livia Drusilla (c.58 B.C.-29 A.D.). Octavian (63 B.C-14 A.D.), the grand nephew and adopted son of Julius Caesar (100-45 B.C.), who later became the emperor Augustus, encountered Nero and Livia at Rome in late 39 B.C. and fell instantly in love with the beautiful Livia, who was then only 19. Though Livia was pregnant with her second child, Nero Claudius Drusus (38-9 B.C.), Octavian took her way from her husband and later that year she was formally divorced and married to Octavian. The children stayed with their father until his death a short time later, after which they were reunited with their mother in Octavian's household, where they were raised with Augustus's daughter Julia (39 B.C.-14 A.D.) by his first wife Scribonia.

Livia's elder son by her first husband, Tiberius Claudius Nero, had been born at Rome on 16 November in 42 B.C. and bore the same names as his father; he was thus Octavian's stepson.

The circumstances of Tiberius's life are given in detail by Gaius Suetonius (69-after 122 A.D.) in his book, *The Lives of the Twelve Caesars.*[1] Here it will suffice to mention that from about 20 B.C. to 6 B.C. Tiberius pursued a strenuous and successful military career as a field general of the Roman army. But in the latter year, since it appeared that Augustus's grandsons would succeed him, Tiberius retired to the Island of Rhodes, where he stayed until 2 A.D., ac-

[1]A convenient source is *The Lives of the Twelve Caesars by Suetonius*, translated by Joseph Gavorse (New York: Modern Library, 1931)

quiring a thorough knowledge of Greek culture and the Greek language.

During his stay in Rhodes he met the learned Alexandrian scholar and astrologer Thrasyllus (d. 36), who cast his horoscope and promised him a great future, which at the time did not seem likely. But in 1 A.D., urged on by Livia's repeated entreaties, Augustus, with some reluctance, began to consider Tiberius as his eventual successor. And when in 2 A.D. a ship arrived in Rhodes summoning Tiberius back to Rome,[1] he became a firm believer in astrology. He formed a lasting friendship with Thrasyllus, and retained Thrasyllus as his personal astrologer for the rest of his life. He also obtained Roman citizenship for his friend, and thereafter the astrologer was known as Tiberius Claudius Thrasyllus.

In 4 A.D. Augustus formally adopted Tiberius and granted him special powers for 10 years. These were renewed in the year 13 and extended in scope, so that until the death of Augustus on 19 August 14 A.D., Tiberius was practically joint-emperor. As emperor, Tiberius was an effective ruler, although he was said to be rather slow and sometimes indecisive or even wavering in dealing with problems that arose. For a time, he maintained good relations with the Senate, spurning grand titles and insisting that he was merely the first (*princeps*) among equals; and he was initially tolerant of mild criticism from the senators. But as he grew older, he became increasingly suspicious of threats to his power or to his life, and he cruelly punished anyone whom he suspected of treachery.

Tiring of the political activities and incessant intrigues of the capitol, in the year 20 he left Rome for Campania, and after residing there for a short time, he settled permanently on the Island of Capri, making only two short trips away from his villa there, on the last of which he was suddenly taken ill, and he died at the home of

[1]As we shall see from the chart below, in 2 A.D. the transiting Uranus was conjunct his Jupiter in the 7th house and Saturn had come to the MC and to the opposition of its natal place, thus indicating an unexpected and significant change for the better in his status.

his friend Lucullus at Misenum (now Miseno) on 13 March 37 at the age of 77.

In addition to a detailed account of Tiberius's military and political career, Suetonius describes in lurid detail Tiberius's sexual improprieties that he indulged in at his villa on Capri when he was already past the age of sixty.

From an astrological point of view, Tiberius was a mixture of Libra and Scorpio, and the Libra influence was gradually overshadowed by the Scorpio influence.

Now we come to the question of his horoscope. Neither Suetonius nor any of the other ancient authorities give Tiberius's birth time, but through a curious circumstance I believe we can determine it.

When Julius Caesar issued his decree in 46 B.C. reforming the Roman calendar, he instructed Leap Year to take place by adding one day in February every four years, rather than by adding a short month in February every other year as had been the irregularly followed custom in the past.[1] His decree contained the Latin words *quarto quoque anno*, which mean 'also every fourth year'. Unfortunately, Romans were accustomed to counting both ends of an interval, so the officials interpreted this to mean what we would call 'every third year', and therefore they celebrated Leap Year every *third* year for twelve successive times until 9 B.C. Thereafter, someone pointed out the error to Augustus, and he decreed that to correct the error no more Leap Years should be observed until the year we call 8 A.D., when the correct 4-year cycle was commenced that has lasted every since.[2]

---

[1]Prior to Caesar's reform in 46 B.C., the Roman calendar had only 355 days (many of the months had fewer days than they have now), so after two of those years it was 20 or 21 days behind. To keep the calendar more or less in line with the seasons, the authorities would add a short month of 22 or 23 days after the 23rd of February every other year. But they sometimes arbitrarily omitted the intercalation.

[2]See the discussion by F. K. Ginzel, *Handbuch der mathematischen und technichsen Chronologie* (Leipzig: J. C. Hinrichs, 1906-14. 3 vols.), vol. 2, "Zeitrechnung der Römer." I have made a table of the deviations of the

Due to this error in the Roman calendar in the years from 42 B.C. to 4 A.D., a recorded date in the Roman Transition Calendar (as we may call it) can deviate from the corresponding date in the Julian Proleptic (astronomical) Calendar by as much as 3 days. In the year 42 B.C. the error was +1 day after February. And since Tiberius was born on 16 November 42 B.C. in the Roman Transition Calendar, the corresponding date in the Julian Proleptic calendar was 17 November.[1]

It is known that a coin was issued with Tiberius's portrait on one side and the sign of Libra on the other side. Thus he was known to have had the Moon in Libra,[2] and this is also mentioned (without his name) in Manilius's *Astronomica.*[3] We can therefore inquire where the Moon was on 17 November 42 B.C. And we find that at 6 A.M. LAT at Rome its tropical longitude was 25°28′ Libra. But the astrologers of the first century B.C. were using the Alexandrian fixed zodiac, which in the year 42 B.C. was about 6°30′ ahead of the tropical zodiac.[4] Consequently, at 6 A.M. LAT the Moon's longitude in the Alexandrian zodiac would have been about 1°58′ Scorpio. Therefore Tiberius must have been born before 6:00 A.M. LAT, and in fact, if we calculate the Moon's longitude at Rome at midnight LAT we find that it was 21°43′ Libra; and if we then add 6°30′ to it, we have 28°13′ Libra for its longitude in the Alexandrian zodiac. Its motion was about 15° per day,

---

Roman Transition Calendar from the Julian Proleptic Calendar.

[1]Not on 15 November as I stated in the first edition of my book, *A History of Horoscopic Astrology* (Tempe, Az.: A.F.A., 1996), p. 24, Note 46 (where I mistakenly subtracted the correction of one day instead of adding it). This error was corrected in the second edition (2006).

[2]The ancients noted a person's Moon sign, rather than his Sun's sign as we do today.

[3]See A. E. Housman's edition, *M. Manilii Astronomicon* (Cambridge: Cambridge University Press, 1937. 2nd ed. 5 vols.), vol. 1, especially the preface, pp. lxix-lxxi, and the Addenda, pp. 93-96.

[4] See my paper, "The Classical Zodiac" in the A.F.A. Journal of Research 7, No. 2 (1995). (It should be noted that the Alexandrian Fixed Zodiac differed from the so-called "Sidereal Zodiac" of Cyril Fagan and his associates by about 3°.)

so for the astrologers it would have remained in Libra in the Alexandrian zodiac until about 4:26 AM LAT.[1] During that interval, the ASC would have varied from 29 Leo to 21 Libra in the tropical zodiac or from about 5 Virgo to 28 Libra in the Alexandrian zodiac; hence, Tiberius must have had either Virgo or Libra rising. But a Virgo rising chart on that morning would have been weak, so either Thrasyllus was given a definite time, such as the 9th hour of the night, or else he was given a time such as "a few hours after midnight," and he saw that a time around the 9th hour would give a Libra ascendant.

These calculations of course assume that the astrologers of that time were able to calculate the Moon's position and the ASC degree as precisely as we are able to calculate them today. They weren't. The Moon's longitude may have been plus or minus 1 degree from its true value. And their method of calculating the ASC degree only yielded an approximation. The usual method of calculating it was to multiply the hour of birth by 15 and then cast that many degrees from the longitude of the Sun by day or from the longitude of its opposition by night. For example, if we assume that Tiberius was born at 3 A.M., which was the 9th hour of the night, then 9 times 15 equals 135°, and if we add that to the opposition of the Sun's longitude in the Alexandrian zodiac, we have 135° plus 29°01′ Taurus, which equals 14°01′ Libra for the ASC degree.

At 3:00 A.M. LAT the tropical longitude of the Moon would have been 23°28′, which, with 6°30′ added, would have been 29°58′ in the Alexandrian zodiac. So, it would have been in 29 Libra as Thrasyllus would have considered it.[2]

Thrasyllus then would likely have obtained a chart with Libra rising and the Moon and Venus rising in that sign. And since

---

[1]The ancients did not have a zero; consequently, the beginning of a sign was understood to be at longitude 1°00′ ("the 1st degree") and the end at longitude 30°59′ ("the 30th degree").

[2]And even if the Moon had been mistakenly calculated a degree further along, it would have been in what he would have considered to be the 30th degree of Libra.

Thrasyllus knew that Tiberius was a very important person, even though he was a little out of favor at the moment, he must have felt certain that the Libra ASC suited him.[1] It certainly was a powerful chart with 4 angular planets, the ASC ruler Venus in its own sign Libra in the ASC and the Light of the Time (the Moon) also rising in the ASC.

But for us, using the tropical zodiac with Placidus house division, the ASC was 4 Libra, which also puts 4 planets in angular houses. The ASC indicates the instinctive nature of the individual, aside from his thoughts, which are indicated by the Moon, and his internal standard of right and wrong, which is indicated by the Sun.[2] In this case both the ASC and the Moon are in Libra, which indicates that Tiberius's instinctive actions and his thoughts harmonized. From the description given by Suetonius, it seems to me that the Libra influence is strong. Particularly in his younger years he was inclined to follow the rules, even to the extent of being pedantically precise; he treated others fairly, and he sometimes had difficulty in judging the proper course of action, even giving the impression of being indecisive or of occasionally wavering between two alternatives. But his Sun and Mercury were in Scorpio, so his internal censor was inclined to personal secrecy and suspicion of the motives of others. And in his middle and later life that Scorpio influence came strongly to the fore, so that if he suspected someone, he thought it only fair to deal with him harshly.

Accordingly, here is a chart with the planets Uranus, Neptune, and Pluto included. It is set for 3:00 A.M. LAT with Placidus cusps. Note particularly the Venus-Uranus conjunction, with both planets in the same sign with the Moon. This undoubtedly accounts for Tiberius's aberrant sexual behavior in late life, aggravated as it was by the partile opposition of Jupiter and the close squares of Saturn and Pluto to Venus.

---

[1]He would have known that Rome itself was said to have been founded when the Moon was in Libra.

[2]See my paper, "An Astrological theory of Personality" in the A.F.A. Journal of Research 12 (Winter 2000).

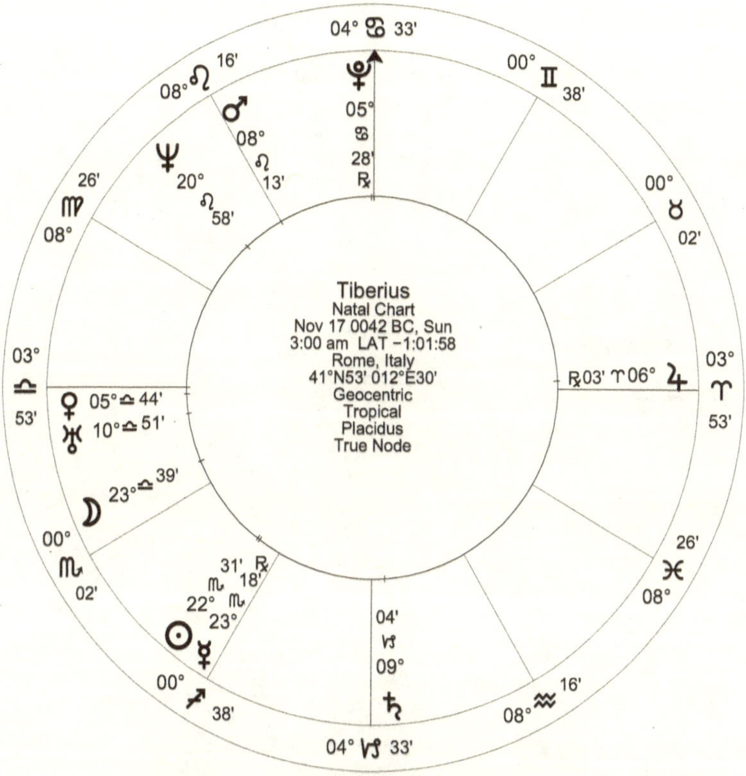

Tiberius
Rome 41N53 12E30 Placidus
17 November 42 B.C. 3 A.M. LAT

We may note that Tiberius's Saturn was conjunct Augustus's Moon, so that he must have seemed like a grim and burdensome person to Augustus; and that since Augustus's Moon was in Capricorn square Tiberius's Moon in Libra, the two men naturally did not think alike, and they consequently disliked each other. Augustus's tropical chart was like this:

We can also note that Tiberius automatically became emperor upon the death of his adoptive father Augustus on 19 August 14.[1]

[1]Practically speaking, as Augustus's designated heir, Tiberius was em-

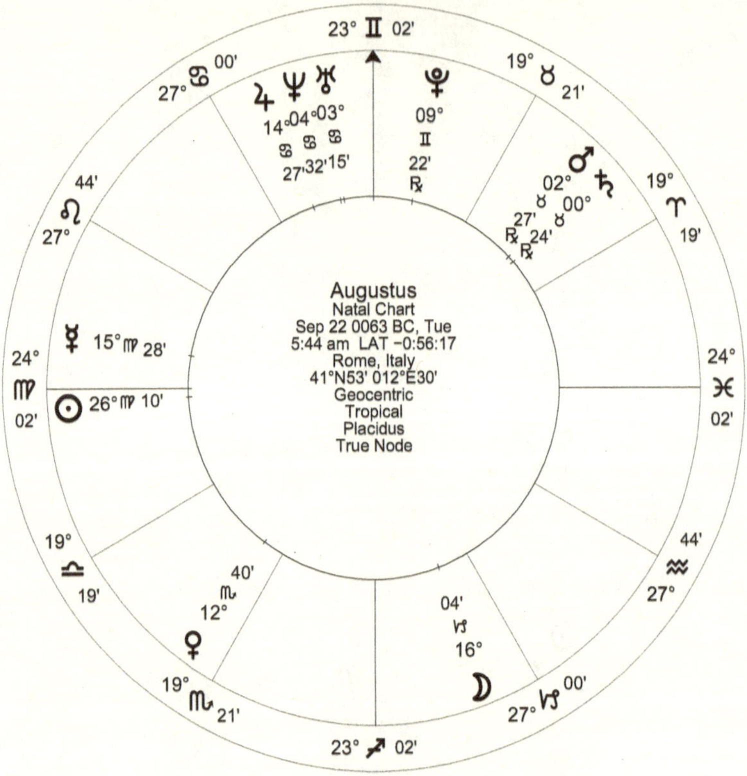

Augustus
Rome 41N53 12E30
22 September 63 B.C. 5:44 A.M. LAT

At that time the transiting Mars was at 14 Leo, near its natal position, and Saturn was at 23 Scorpio in partile conjunction with Tiberius's natal Sun. Lewi describes the signification of these transits as follows.[1] For Mars: "A period when new starts should be made, new plans put into execution, new ideas promoted" . . .

---

peror from that day, although he was only officially acclaimed emperor by the Senate a month later on 17 September.

[1] Grant Lewi, *Astrology for the Millions* (New York: Doubleday, Doran & Co., 1940. 1st ed.; Saint Paul: Llewellyn Publications, 1977. 5th ed. revised).

You are starting a two-year cycle" . . . "you will find that these recurrent two-year periods marked with great consistency periods of important decisions, developments, activity and changes." And for Saturn: "This position brings you the results of what you have accomplished in the past. Under it *you get what you have worked for and what you deserve.* . . . "In its best manifestation this period sets the crown of achievement on the deserving head of effort; and thus it may mark a giddy period of riding the crest of popularity, success and accomplishment."

Both of these indications were certainly appropriate for a man who at that time in his life became the Ruler of the Western World!

Finally, it is interesting to note that Tiberius's corrected birthday, 17 November, is the same as that of the emperor Vespasian, who was born exactly fifty years later on 17 November 9 A.D.

CLAUDIUS PTOLEMY

by James Herschel Holden, M.A., FAFA

Abstract: An account of the professional career of Claudius Ptolemy, the 2nd century writer on astronomy, astrology, and geography, with an explanation of why Ptolemy wrote his books and what happened to them after they were written. The usual assumption that they became available immediately after they were written is refuted.

Claudius Ptolemy is generally estimated to have lived from about 100 to 178 A.D.[1] and to have flourished between the years 124 and 141. The latter date may possibly be extended somewhat by the so-called "Canobic Inscription" that contains some of Ptolemy's astronomical parameters,[2] since it ends with the statement

[1]What little is known of Ptolemy's life is summarized by Otto Neugebauer in *A History of Ancient Mathematics and Astronomy* (New York Heidelberg Berlin: Springer Verlag, 1975. 3 vols.), vol 1, pp. 834-835. The estimated birth year of 100 may not be far from the truth, but there is no evidence at all for how long he lived after the middle 140's, and in fact virtually nothing is known of his life. His given name Claudius implies that he was a Roman citizen; and his surname Ptolemy might imply that he was a native of the town of Ptolemais. Indeed a 14th century writer says that he was born at Ptolemais Hermeiou, a little town near the modern Akhmîm on the upper Nile. But there were other towns named Ptolemais, e.g. the modern Tolmeta (Tulmaythah), 70 mi. E of Benghazi, Libya; and the "naval station" Ptolemais south of the modern El Faiyûm. Or, it is even possible that the name (but not necessarily the person) may have been related in some way to the famous Ptolemies, Kings of Egypt.

[2]See Neugebauer, *op. cit.*, vol. 2, pp. 913-917. But Robert R. Newton, *The Origins of Ptolemy's Parameters* (The Center for Archaeoastronomy, University of Maryland, 1982), p. 4, cites a statement by G. J. Toomer to the effect that he doubts that the Canobic Inscription is au-

that it was written in the 10th year of the emperor Antoninus, which corresponds to about 146 A.D.

Ptolemy's astronomical treatise, the *Syntaxis*, later called the *Almagest*, eventually became the subject of a now fragmentary commentary by Pappus, who seems to have lived around 320 A.D., and an extensive commentary by Theon of Alexandria who lived around 360 A.D. Theon also prepared an edition of Ptolemy's *Handy Tables*.[1] But, so far as is known, there is no indication that Ptolemy's astronomical works were known to the general public from the time of their completion in the mid-2nd century until the 4th century!

A similar testimony to the knowledge, or rather the non-knowledge, of Ptolemy's works in the intervening century and a half is that his astrological treatise, the *Tetrabiblos*, seems to have been first mentioned by the philosopher Porphyry, who wrote an *Introduction to the Tetrabiblos* around the year 295 A.D.[2] But it is particularly significant that the professional astrologer and teacher Vettius Valens,[3] who was living in Alexandria during the years 160 to about 175, had never heard of Ptolemy or any of his books![4]

---

thentic; and if it is not, we are back to the early 140s. However, Toomer seems to have had second thoughts and later suggested that it might be a work of Ptolemy composed prior to the *Almagest*, since a few of the parameters are different.

[1]See the discussion by Neugebauer, *op. cit.*, vol. 2, pp. 965 ff.

[2]The Roman senator and amateur astrologer Julius Firmicus Maternus, who completed his massive astrological treatise the *Mathesis* around 355 A.D., mentions Ptolemy three times, although he evidently did not have a copy of the *Tetrabiblos*. But twenty years later, the astrological writers Paul of Alexandria (378 A.D.) and the so-called 'Astrologer of the Year 379' did have at least some of Ptolemy's works. And Theon of Alexandria, who flourished around the year 370, had at least the *Syntaxis* and the *Handy Tables*.

[3]See *Vettii Valentis Antiocheni Anthologiae libri novem* ed. by David Pingree (Leipzig: B.G. Teubner, 1986). The Index Auctorum on pp. 456-457 has an entry for Ptolemy, but all the references are to the supplementary Byzantine scholia and florilegia relating to the *Anthology*. The name Ptolemy does not appear anywhere in the main text of the *Anthology*!

[4]Curiously, however, Paul of Alexandria (fl. 378) says, *Introduction*, 1,

Thus, it appears that both Ptolemy's astronomical and astrological works did not become available to the public until the beginning of the 4th century, and the same may probably be said for his *Geography* and his other works. What are we to make of these facts? I will propose a hypothesis.

Ptolemy's works are dedicated to an otherwise unknown Syrus. It seems likely that Syrus was a wealthy individual who desired to include in his personal library works on the sciences. He became acquainted with Ptolemy and hired him to write synopses or summaries of the sciences of astronomy, astrology, geography, etc. When these works had been completed, Syrus placed them in his personal library. And there they remained for a century and a half before they were sold to a local bookseller, after which copies could be obtained by those who were interested.

If indeed Syrus was a wealthy individual, his descendants may have lived successively in his house until around 300 A.D. Then, either the family died out, or the then current descendant decided to get rid of the old books. The contents of the library, or at least the books by Ptolemy, were sold to a local bookseller. Recognizing that there might be a demand for such books, he circulated the information that he had the originals and could supply copies. Local citizens who wished to do so procured copies, and the existence of the books gradually became known to the few persons of the time who were interested in astronomy, astrology, or geography.

If this hypothesis is accepted, it explains Ptolemy's place in 2nd century society. He was one of those rare individuals who had a

---

that in one of his books Apollinarius advocated the solar longitudes of Ptolemy, which would imply that he had read either the *Almagest* or the *Handy Tables* of Ptolemy. And Valens mentions Apollinarius in *Anthology*, vi. 4 in connection with phenomena. But Valens used solar longitudes related to the fixed Alexandrian zodiac, so it is plain that Valens did not find any mention of Ptolemaic longitudes in Apollinarius's book. Apollinarius is also mentioned by Porphyry, *Introduction,* 41, but only in connection with the disposition of the terms. It remains hard to believe that if Ptolemy's books had been available in Alexandria in the second century, Valens would have been unaware of them. So I still believe that they did not begin to become available until around 300 A.D.

strong personal interest in a subject and studied everything he could lay his hands on to perfect his knowledge of it. He was able to acquire copies of astronomical works by older authors, such as Hipparchus. He may even have made a few astronomical observations himself. He was obviously a competent mathematician and a dedicated student. But he was not a professor of astronomy, he was not a diligent observational astronomer, he was not a scientist in the modern sense of the word, and he did not hesitate to exaggerate and even misrepresent his own astronomical activities and those of his predecessors. He was, in short, a competent and dedicated amateur astronomer, mathematician and scholar, and a good expositor, but, as Newton has shown, a plagiarist, and a bit of a liar.

Newton has amply demonstrated that nearly all of Ptolemy's professed "observations" are fraudulent.[1] And his star catalogue is simply that of Hipparchus with a constant added to the longitudes (and with some of the longitudes deliberately altered slightly to disguise their origin). In a later work, Newton examined the astronomical tables in the *Almagest* and concluded that at least some of them were probably copied from tables prepared by earlier writers on astronomy, whose works have not come down to us.[2] And in his astrological work, the *Tetrabiblos*, Ptolemy himself explicitly ac-

[1]See Robert R. Newton, *The Crime of Claudius Ptolemy* (Baltimore and London: The Johns Hopkins University Press, 1977). Curiously enough, the few observations that Ptolemy records that Newton thinks are genuine are ones that he did not use, probably because he saw that they did not agree exactly with the theories that he had adopted from older writers on astronomy.

[2]Robert R. Newton, *The Origins of Ptolemy's Astronomical Tables* (The Center for Archaeoastronomy, University of Maryland, 1985), p. 8. Here, Newton calls particular attention to the fact that the astronomical tables only cover a period of 810 Egyptian years begining 26 February -746 and ending 29 August 82. Newton says: "I can see no reason why an astronomer, in preparing tables of all the important mean motions, would make them stop in his grandparents' time. The only explanation I can think of is that Ptolemy did not prepare these tables. Instead, he simply copied them from some earlier source, without even bothering to do the calculations needed to bring them down to his own time." I agree with Newton.

knowledges that he is summarizing the work of previous writers on that subject, when he says:

> ". . . we shall decline to present the ancient method of prediction, which brings into combination all or most of the stars, because it is manifold or well-night infinite, if one wishes to recount it with accuracy."[1]

Newton discusses his own and others' efforts to understand and possibly legitimize Ptolemy's motivation in writing the *Almagest* the way that he did,[2] and Newton raises the question as to why Ptolemy thought that he could get away with such fraud. Would not his peers have seen through his mistreatment of his predecessors' observations and have realized that he had not made any significant observations of his own? I say, No, they wouldn't! For a simple reason. He had no peers!

Ptolemy was not preparing his books as a contribution to organized science. There was no organized science! He was preparing them under contract for a private individual.[3] And if his work was to go into a private library, then no one but his patron Syrus and Syrus's family and friends would see it. They would certainly not be capable of detecting any fraudulent work. And even if his work should become known to the public, who was there in 2nd century Alexandria (or anywhere else in the Mediterranean world) that would have known more about the subject than Ptolemy himself? Probably no one! There were no observatories. There were no professional astronomers. There were no professors of astronomy.

---

[1]Ptolemy, *Tetrabiblos* ed. and trans. by F.E. Robbins (Loeb Classical Library, 1940), iii.1. And in various places in the *Tetrabiblos* Ptolemy refers to "the writers" or "the ancient" (probably Nechepso), so it is plain that he is summarizing previous books on astrology.

[2]See, for example, the sections entitled "A Pedagogical Motivation for Fabrication, " parts (a) and (b) in R.R. Newton, *The Origin of Ptolemy's Parameters* (The Center for Archaeoastronomy, University of Maryland, 1982), pp. 159 ff.

[3]It is reasonable to assume that Syrus paid a stipulated sum to Ptolemy to write the books. Unless Ptolemy was independently wealthy, he had to have an income to survive.

The few individuals who lectured on "astronomy" in what passed for universities in the second century probably scarcely covered more than the simple facts of astronomy. There were no astronomical journals. And it is doubtful that there were even any equally well-informed amateur astronomers.

In raising his question about Ptolemy's motivation, Newton was perhaps unconsciously thinking about the present-day situation in which any new astronomical work issued either in book form or as an article in a scientific journal is immediately subjected to critical scientific scrutiny the moment it is published. This was not the case in classical antiquity! Possibly the only scholarly subject that would have attracted the attention of any potential critics in those days was philosophy.

Today there are thousands of professional astronomers and perhaps a hundred or more astronomical journals of one sort or another. Any new idea or treatise that appears immediately becomes the subject of scientific discussion. But we live in a world in which "science for science's sake" is an accepted ideal. This was not the case in antiquity. Who, for example, would have had any interest in knowing the motion of the planet Mars? Only an astrologer. He had a definite practical use for that information. No one else did![1]

But weren't there any "true scientists" in the 2nd century who would want to know about such things? No! There were possibly one or two curious amateurs who might have taken an interest in the problem of planetary motion as an intellectual challenge or as an interesting problem in mathematics. In fact, Ptolemy himself was probably just such a person. But people have to make a living.

[1]It is interesting to note that John Kepler's astronomical research was supported financially by the emperor Rudolph II (1552-1612), who was interested in astrology and therefore saw the need for improved planetary tables. (Kepler acknowledged this by using the Emperor's horoscope as the date for the example calculation of solar, lunar, and planetary positions in his *Rudolphine Tables*.) But Rudolph's successors, the emperors Matthias (1557-1619) and Ferdinand II (1578-1637) were not interested in astrology and consequently could see no practical use for more accurate planetary positions, so they quit paying Kepler.

As mentioned above, there were no chairs of mathematical astronomy in the few schools that could even remotely be likened to today's universities. There were no observatories to offer remunerative employment. There were no government grants that could be grabbed to finance research. And perhaps the only private grant that we can find traces of in classical history is the one that Syrus evidently gave to Ptolemy.

Thus we see that Ptolemy was a unique individual, an intelligent man who had the understanding of mathematics, the intellectual curiosity, and the perseverance necessary to search out and survey the existing literature and write expository treatises that summed up the available knowledge of astronomy and geography and some other subjects. He undoubtedly had access to whatever books still remained in the Alexandrian Library, and he may even have known an elderly amateur astronomer who had made a few observations or who had some useful knowledge of the subject.

In some ways, then, Ptolemy's astronomical work may be compared to that of Euclid (fl.c. 300 B.C.), who summarized all of the geometrical knowledge available in his time in a systematic and near perfect format. Or to that of the American prodigy Nathaniel Bowditch (1773-1838), who translated the *Celestial Mechanics* of La Place (1749-1827) and added very detailed explanatory comments (and occasional corrections), so that his translation is a much more informative work than the original. If Laplace's French original had perished and no subsequent astronomical treatises had been written, Bowditch would certainly enjoy a much more exalted place in the history of astronomy. Notwithstanding the fact that he was an insurance man by profession and no more than a self-taught mathematician and an amateur astronomer.[1]

It is interesting to note that modern historians of science have elevated Ptolemy to the same status as that in which he was held by

---

[1]Bowditch, who had made five sea voyages while in his twenties, also published in 1802 *The New American Practical Navigator*, a complete handbook for seamen, that became the "bible" of celestial navigation and had appeared in more than 60 subsequent editions by the middle 1950s.

some astrologers at the end of the classical period. He is in effect the "divine Ptolemy"—a great hero of pure science! As such, when his astronomical work was subjected to a careful mathematical examination by Newton, and when that examination revealed and documented Ptolemy's fraud, the historians of science were aghast! How could anyone have the temerity to challenge the integrity of the greatest scientist of antiquity? No matter that Newton's detailed critique of the "observations," the star catalogue, and the astronomical tables definitively revealed what Ptolemy had actually done (and not done). The scholarly attitude was, "My mind is already made up; don't try to confuse me with the facts!" Surely Newton must have been mistaken, and in any event he could simply be ignored. (He was, after all, only a physicist and not a science historian!)

Science is supposed to be a search for knowledge. It is truly disappointing to see new research disregarded in order to maintain false notions that have become enshrined in the college textbooks. That is precisely the attitude that Galileo encountered from the academics in the 17th century when he began to report the discoveries that he had made with his new telescope. He was scoffed at by the learned professors (and by the Pope).

Admittedly, what I have proposed above is an unprovable hypothesis. But it seems to me that it provides a reasonable explanation of the known facts. It explains why Ptolemy undertook the task of summarizing scientific knowledge in several fields, how he went about it; and why, if he was as great a fraud as Newton says he was—and Newton is undoubtedly correct—he could "get away with it." This does not in any way diminish his credit as an expositor of learned subjects,[1] but it does assert that he was a "science

[1]I cannot, however, agree with Newton's suggestion that the success of Ptolemy's astronomical works has probably deprived us of more authentic astronomical treatises that might otherwise have been preserved. Since, as I have demonstrated, his works did not become available until the 4th century, it seems more likely that very little in the way of older books would still have been available by that time. Hence, if Ptolemy's books had not been written, and by a lucky chance eventually made avail-

writer," a learned theoretician at best, but not a competent observational astronomer.[1] And by today's standards of scientific probity, he can no longer be considered to be a "hero of pure science."

Nor was Ptolemy either the inventor of astrology or a practicing astrologer. His book the *Tetrabiblos* omits much of the Greek astrology that was known in his day. It contains no example horoscopes. It mentions no astrologer by name. Most of it is obviously a compilation from earlier books that are now lost. Perhaps the only original parts are the philosophical comments in Book 1 and the explanation of what would now be called Placidian Primary Directions in Book 3. But it is all very neatly arranged. Like his astronomical works, when the *Tetrabiblos* became available to the astrological community in the early fourth century, there was not much else to choose from—most of the older books were already lost or were very difficult to come by. And since Ptolemy's astronomical tables became available at the same time, their obvious improvement over the older tables provided what appeared to be very accurate solar, lunar, and planetary positions[2] and rising

---

able to the public, we might very well know little or nothing of the work of the older astronomers, since it seems reasonable to suppose that Ptolemy had access in the 2nd century to books that were no longer in existence by the 4th century. I believe that it is better that we have some notice of older astronomers and their theories, tables, and observations (even if they have been unscrupulously tampered with), than to have no notice of them at all.

[1]And Ptolemy's egregious error in reducing the circumference of the Earth from Eratosthenes's earlier and nearly correct figure of 25,200 miles to only 18,000 miles does not speak well for his skill as a geographer. Still, by drastically reducing the estimated distance westward from Spain to the Orient, he may well have encouraged Columbus to undertake his epic voyage of 1492.

[2]I say "appeared" because Ptolemy had adopted an erroneous value of the general precession in longitude (36" per year rather than the nearly 50" that was the true value in his time). Consequently, by the middle of the fourth century, when his tables became available, his "tropical" longitudes (aside from other errors) were already 1.8° in error (which no one seems to have noticed at the time). Astrologers have generally trusted astronomers to provide them with accurate tables and ephemerides. But even the late fourth century astronomer Theon did not notice the error in

times. This established his reputation as a learned astronomer, and that spilled over onto his astrological work. Small wonder, then, that Ptolemy acquired an equally great reputation as a master astrologer in late antiquity.

the Ptolemaic longitudes.

THE PURPOSE OF THE TREPIDATION OF THE EQUINOXES

by James Herschel Holden, M.A. FAFA

Abstract: An explanation of the so-called Trepidation of the Equinoxes.

In the 4th century the astronomer Theon of Alexandria included a curious passage in his edition of Ptolemy's *Handy Tables* that described a supposed oscillation of the equinoxes that he attributed to "old astrologers." This puzzled astronomers ever after. In the present paper, I explain that it offered a means of calculating the difference between the tropical equinox introduced by Hipparchus and the equinox of the fixed zodiac used by the Alexandrian astrologers. To use the Hindu term, it was an *ayanâmśa*.

In the *Handy Tables* of Theon of Alexandria (4th century), we find the curious chapter, "On the Tropic," which has puzzled astronomers ever since. Here is the text of the chapter:

On the Conversion

The old astrologers pretending, based on certain conjectures, that the tropical points, having been advanced towards the east by 8 degrees since a certain epoch of time, would be eventually returned to where they were [at that time], a thing that does not seem true to Ptolemy,[1] in view of the fact that, despite that pretension, the calculations made by the tables are in agreement with the observations made by instruments; we should, therefore, not adopt this correction. However, we are going to set forth the method that these astrologers follow in their calculations.

[1]This is apparently an *argumentum ex silentio*, since Ptolemy never mentions such a conjecture in the *Almagest*.

They take the 128 years that preceded the Era of Augustus as being the quantity of time that that movement of 8 degrees [began] towards the points following, and at the end of which it will have commenced its return into its preceding state; and adding to that the 313 years that have elapsed from the reign of Augustus down to the reign of Diocletian, along with the given number of years since Diocletian, they take the place of this sum by reason of a degree of movement of this place in 80 years; then, subtracting the quotient of this division from 8 degrees, they add the remainder, as marking the degree down to which the tropic points have then advanced, to the degrees obtained by the above-mentioned calculations of the place of the sun, the moon, and the five planets.[1]

In other words, according to Theon, to the longitudes of the Sun, Moon, and Planets derived from the *Handy Tables*, the "old astrologers" were accustomed to add a correction based on the motion of the tropic points. This correction is stated to have been in the form of what Neugebauer has termed a "zigzag function" between the limits of +8° and 0° with a period of 640 Egyptian years and an epoch of 2 October -157. In other words, the correction would decrease from +8° towards 0° at the rate of 45″ per Egyptian year, and then it would reverse itself and increase at the same rate until it again reached +8°, after which the cycle would begin again.[2]

Theon obviously did not understand the reason for such a correction. And no astronomer after him did either. Yet, the reason is plain enough if you think about it. Notice that it was the "old astrologers" who used this correction. Why would they have done so? For a very simple reason! Horoscopic astrology was invented by Greek-speaking Alexandrians in the early part of the 2nd cen-

[1]Translated from the French paraphrase by the Abbé Nicolas Halma, *Commentaire de Théon d'Alexandrie sur le livre iii de l'Almageste de Ptolemée* (Paris: A. Bobée, 1822), p. 53. As a title, Halma translates the Greek *tropês* 'tropic' as 'conversion', which is incorrect, but it describes the subject of the chapter fairly well.

[2]It is usually referred to in the scholarly literature as the *Trepidation of the Equinoxes*.

tury B.C. They used a fixed zodiac. They prepared astronomical tables that gave the positions of the Sun, Moon, and planets referred to that fixed zodiac. Those tables continued in use for centuries. But when Hipparchus discovered the precession of the equinoxes in the latter part of the 2nd century B.C., he prepared tables of the Sun and Moon referred to the tropical zodiac. It seems likely that the few astrologers who obtained copies of those tables presumed that they were more accurate than the older tables they had been using. However, if they used them, a problem immediately arose. The solar and lunar positions yielded by Hipparchus's tables differed by several degrees from those yielded by the older tables.

Hence, there was a need for a correction factor to reduce Hipparchus's tropical longitudes to longitudes referred to the Alexandrian fixed zodiac.[1] The difference in longitudes was evidently assumed to be +8° in the year -157 because that was the longitude of the Sun in the fixed Alexandrian zodiac at the time of the vernal equinox in that year. Hipparchus had determined that the precession amounted to 45″ per year. The correction factor, therefore, simply corrected the tropical longitudes of the Hipparchean tables to longitudes in the Alexandrian fixed zodiac.

And since Ptolemy later adopted Hipparchus's tropical longitudes, it might be supposed, as Theon evidently did,[2] that the correction could be applied to longitudes calculated from the *Almagest* tables or from the *Handy Tables*. However, since Ptolemy had not adopted Hipparchus's precessional motion of 45″ per year but had instead adopted a motion of only 36″ per year, the correction

[1]Greek astrologers continued to use positions referred to the Alexandrian fixed zodiac until Ptolemy's *Handy Tables* became readily available in the 4th century. Then, they switched from positions referred to the old Alexandrian fixed zodiac to positions referred to the "tropical" zodiac of Ptolemy's tables. That change in the reference zodiac from fixed to tropical has persisted to this day in Western astrology.

[2]And perhaps Theon knew of some astrologers in his own time who were applying the correction to the positions from Ptolemaic tables. But this does not seem likely, since he says it was the "old astrologers" who used the correction.

was thrown off by 9″ per year. To adapt the "tropical" longitudes of the *Almagest* and the *Handy Tables* to the Alexandrian fixed zodiac, the formula would need to have been changed to +7°37.5′ - 1° per hundred years measured from the epoch -128.[1] But since Theon didn't understand the reason for the correction, he would not have realized that it needed to be modified if it was to be applied to Ptolemy's tables.[2]

So far, so good! The purpose of the correction is now clear. But those who had devised the correction evidently thought for some reason that the motion of the equinoxes was cyclic. Therefore, they specified that the correction should only go to zero and then reverse itself. This seems absurd to us, but we should remember that in the 2nd century B.C. no one knew why the equinoxes moved, so it was not unreasonable to suppose that their motion might be cyclic. And since the year of reversal would not arrive until 482 A.D., it was too far in the future to worry about. In fact, the classical period had very nearly come to a close before that date was reached.

The Arabian astronomers of the 8th and 9th centuries could easily see the fallacy of the cyclic motion of the correction, so it was obvious to them, as it has been to nearly everyone since that time that the *oscillatory part* of the correction was false.[3] However, during the whole of the classical period, the correction was valid for the purpose for which it had been designed!

---

[1]The epoch would need to be changed because Ptolemy's "tropical" longitudes approximately coincided with Hipparchus's longitudes in the year -128. And the change from 1° per 80 years to 1° per 100 years would reflect the change in the precessional motion from 45″ per year to 36″ per year.

[2]And if any astrologers were applying the correction to Ptolemaic positions in the 4th century, they would probably not have been aware that Ptolemy's location and motion of the equinox differed from Hipparchus's values.

[3]However, it was not completely obvious to Copernicus, who reduced the Trepidation of the Equinoxes to a cyclic variation of the rate of precession. See J. L. E. Dreyer, *A History of Astronomy from Thales to Kepler* (New York: Dover Publications, 1953. 2nd ed.), pp. 329-330.

As was said, Theon did not understand the purpose of the correction. He may not have been interested in astrology. Or, if he was, he may not have been aware that the "old astrologers" had used solar, lunar, and planetary positions referred to a fixed zodiac. By his time, the nominal value of the correction had decreased to about +1.5°, and even if he had seen any of the older tables, he may have simply thought that they were less accurate than Ptolemy's tables, which they were, and he may not have realized that not only was their theoretical basis less accurate, but that they also gave positions referred to a fixed zodiac.

Interestingly, Pingree[1] has stated that Greek astrology passed to the Hindus in the 2nd century A.D., when one or more Greek astrological treatises were translated into Sanskrit. These treatises were based upon the Alexandrian fixed zodiac, and they may possible have contained a statement of the correction that Theon mentioned. For, if we calculate the interval from the year -157 to the year +499, we have 656 years. According to the theory, the correction should have gone to zero and then have begun to increase for 16 years, which would have given it a value of +0°12′; or, if the oscillatory feature of the correction had been ignored, a value of -0°12′. Either way, the correction was close to zero in the year 499 A.D.

It was in the year 499 that the Hindu astronomer Âryabhaṭa determined that the Sun's mean longitude had attained the longitude of 0° at Ujjayin (75E46) at either midnight 20/21 March or at 6 AM 21 March.[2] He adopted its position on that date as the zero point of a fixed zodiac.[3] Thereafter, the difference between that fixed point (or one close to it) and

---

[1]David Pingree, *The Yavanajâtaka of Sphujidhvaja* (Cambridge, Mass: Harvard University Press, 1978. 2 vols.)

[2] At midnight 20-21 March 499 the actual mean longitude of the Sun as determined from modern data was 359.26°, and 6 hours later it was 359.51°. His adopted length of the sidereal year was 365.25875.

[3]See Roger Billard, *L'Astronomie Indienne* (Paris: École Française d'Extrême-Orient, 1971), pp. 79-80. And note that, unlike their Western counterparts, Indian astrologers continue to use longitudes referred to a fixed zodiac.

the moving vernal equinox was calculated and termed *ayanâmśa*, a practice that is still followed by Indian astrologers to this day.[1]

Thus, we see that the type of correction first mentioned in the West by Theon of Alexandria survives in India today, where, just as it did among the "old astrologers" of whom Theon speaks, it serves to indicate the difference in longitude between tropical positions and those referred to a particular fixed zodiac. It is in fact nothing more than the *ayanâmśa* that was used by the old Greek astrologers when they wanted to refer the "tropical" longitudes[2] from Hipparchus's tables to the fixed Alexandrian zodiac.

---

[1]Modern ephemerides are of course calculated from theories that yield tropical longitudes, so Indian astrologers are obliged to apply the *ayanâmśa* correction to those tropical longitudes in order to obtain positions referred to one of the fixed zodiacs that is used in Hindu astrology. *The Indian Ephemeris and Nautical Almanac* issued annually by the Indian Government contains a table of the "True Ayanâmśa" for every third day of the year. (The "true" *ayanâmśa* is the mean *ayanâmśa* + the nutation in longitude.)

[2]But since Ptolemy started out with the equinox in the wrong place and used a deficient value of the precession, his longitudes were only approximately tropical. Hence, his "tropical" longitudes were not the same as those of Hipparchus, as mentioned above.

BOOK REVIEWS

Reviewed by James Herschel Holden, M.A., FAFA

Christino, Karen, *Foreseeing the Future* (Amherst. MA: One Reed Publs., 2002)

A biography of the famous American astrologer Evangeline Adams. Christino has researched Adams's early life and her career as a professional astrologer. Illustrated with many pictures. An engaging account of perhaps the best known American astrologer.

What Evangeline Adams Knew (Brooklyn Heights, NY: Stella Mira Books, 2004)

A sequel to the preceding book. Here Christino adds many additional interesting details about Evangeline Adams and her astrological career. Illustrated.

Hübner, Wolfgang. Ph.D., *Grade und Gradbezirke der Tierkreiszeichen* (Stuttgart & Leipzig: B. G. Teubner, 1995. 2 vols.)

This is a scholarly work in the German language on the significations of the individual degrees and degree areas (such as the *terms*) of the zodiac. Prof. Hübner assembles all the ancient Latin and Greek sources, as well as his edition of the early 14th century translation of a portion of the *Liber Hermetis* into the Picard dialect of French, and he provides a useful discussion of their inter-relations. For those interested in Classical Astrology, Hübner's work is a valuable resource for the detailed structure of the zodiac.

Knipe, Humphrey, *The Nero Prediction* (Los Angeles: Process, 2005)

A fascinating historical novel dealing with the life of the Emperor Nero as seen by the slave Epaphroditus, who became Nero's confidant and ultimately his executioner. The very strong influence of astrology on the lives of the emperor, his relatives, and other Roman aristocrats is carefully described. Notable astrologers, such as Thrasyllus, his son Balbillus, and one who called himself Ptolemy Seleucus, are notable players in the tense drama surrounding Nero. Readers with an interest in astrology will particularly enjoy this book.

McMinn, David, *Market Timing by the Number 56* (Blue Knob, Australia: Twin Palms Publ., 2002)

Market Timing by the Moon and the Sun (Blue Knob, Australia: Twin Palms Publ., 2002)

Market Timing by the Moon Sun Research 2006-2009 (Blue Knob, Australia: Twin Palms Publ., 2009)

These three volumes summarize McMinn's lengthy and detailed studies of market cycles. Those who are interested in financial astrology will find them particularly interesting. They contain very extensive tables of market data over a long period of years, thus providing a remarkable treasure trove of material for additional studies of market behavior.

More recently, from his study of market cycles McMinn predicted a "financial panic" to occur in October 2008. We can certainly see the accuracy of that prediction! And he currently expects another drop in the markets in 2010.

These books may be obtained directly from the author. His e-mail address is mcminn56@yahoo.com.

Avelar, Helena and Luís Ribeiro, *Astrologia Real* (Cascais, Portugal: Editora Pergaminho, 2003)

The authors of this book in the Portugese language are Portugese astrologers. They have very carefully researched all available historical records to establish the birthdates and birthtimes of the kings and queens of Portugal from the earliest times down to the last king, Manuel (1889-1932) in the 20th century. Horoscopes are given along with astrological commentaries, and the book is illustrated with portraits of the kings and queens.

An excellent resource of royal horoscopes!

This book may be obtained directly from Mr. Ribeiro. His e-mail address is contacto@academiadeastrologia.com.

Oestmann, Günther, Ph.D., *Heinrich Rantzau und die Astrologie* (Brunswick, Germany: Braunschweigischen Landesmuseum, 2004. 4to)

Prof. Oestmann has researched the life and career of the Danish Scholar, Statesman, and Astrologer, Count Rantzau (1526-1598). The book is in large format on plate paper and beautifully illustrated. It contains numerous excerpts from Rantzau's letters and books and from books by some of his contemporaries. Astrologers who can read German (and Latin) will be fascinated by the extensive detail on the life and times of this notable astrological writer. It is a valuable contribution to the history of astrology in the 16th century. This book can be ordered from amazon.com.

Riske, Kris Brandt, M.A., PMAFA, *Astrometeorology* (Tempe, AZ: A.F.A., Inc., 1997)

This book is the result of a careful and extensive study of notable storms that occurred in the western Hemisphere during the 20th century. Riske has determined the astrological indications that preceded these storms and deduced rules for making future predictions. Those interested in astrometeorology will find this book a useful guide and a valuable data resource.

Llewellyn's Complete Book of Astrology: The easy Way to Learn Astrology. (St. Paul, MN: Llewellyn Publs., 2007)

A complete and comprehensive guide to natal astrology.

Mapping Your Future (St. Paul, Mn: Llewellyn Publs., 2004)

Riske has prepared a popular guide to interpreting a natal horoscope. The book comes with a CD-ROM disk that enables the reader to cast his own horoscope and explore his future prospects.

Mapping Your Money (St. Paul, MN: Llewellyn Publs., 2005)

In this book Riske explains the features of a horoscope that particularly relate to money prospects. The book comes with a CD-ROM disk that enables the reader to cast his own horoscope and explore his financial prospects.

Stuckrad, Kocku von, Ph.D., *Geschichte der Astrologie* (Munich: C. H. Beck oHG, 2003)

An excellent cultural history of astrology in the German language by Prof. Stuckrad of the Philosophy Department of the University of Amsterdam. An English translation is said to be in preparation.

This book can be ordered from the German branch of amazon.com.

OBITUARIES

Doris Chase Doane, B.A. LPMAFA FAFA MPAI (1913-2005)

She was born in Mansfield, Massachusetts 4 Apr 1913 1:57 AM EST, the eldest of five children of Willis and Rose Chase. She became interested in astrology in 1936, and entered the field as a teacher in 1938. Having relocated to California, she studied at UCLA and graduated with a degree in psychology in 1944. That same year she received the Hermetician Certificate for having passed the 21 Brotherhood of Light courses and began to teach classes at the Church of Light in Los Angeles. She married the Rev. Edward Doane (1892-1970) of that organization and became an ordained Minister herself in 1946.

Mrs. Doane was closely associated with the work of the Church of Light from the early 1940's until 1972, when she moved away from Los Angeles. During that period she served as its Secretary (1949), Director of Astrological Research (1952), Editor of the Church of Light *Quarterly* (1951-1969), and taught classes (1944-1968). She began to write for publication in 1945, and since that time more than 1500 of her articles and books have been published. A number of her books have become standard authorities in their fields, on such diverse subjects as the *Index to the Brotherhood of Light Lessons*, the series on Time Changes, Tarot Card reading (written jointly with King Keyes), and *Horoscopes of the U.S. Presidents*.

Rev. Edward Doane passed away in 1970, and Mrs. Doane married the astrologer John Lawson Ahern (1922-2004) the following year. She served as President of the AFA from 1979 to 2004. She was also Co-Founder (1968) and President Emeritus of Professional Astrologers, Inc., a charter member of the Athena Astrological Society, a long-time member of the Astrologers Guild of America, and a member of several other astrological organizations. In 1988 she was awarded fellowship in the AFA. Mrs. Doane traveled widely, lecturing and conducting seminars and workshops, and was the recipient of many awards from both American and foreign astrological groups. Throughout her career, she was a strong advocate of professionalism and certification in astrological practice. Mrs. Doane passed away in California on 6 June 2005.

Many of her books are available from the AFA Book Department.

David Edwin Pingree, Ph.D. (1933-2005)

Professor Pingree was born on 2 January 1933 at New Haven, Connecticut. He was not well known to astrologers, but he should have been. He was especially interested in the history of mathematics and headed the Department of the History of Mathematics at Brown University from 1970 until the time of his death. He had an almost unrivalled knowledge of foreign languages, including Latin, Greek, Arabic, and Sanskrit. He realized early in his career that since astrologers depended upon astronomers to supply the planetary and solar tables that they needed, it would be useful to study the works of the early astrologers. To this end, he edited several early astrological texts and wrote many papers on early astrology.

Nevertheless, his main interest was the early history of mathematical astronomy. During his lifetime he produced important translations and studies of both astrology and astronomy. He had also prepared manuscript editions of some more astrological works, but unfortunately he had not made arrangements for their printing, so it is uncertain whether they will ever be published. But fortunately it was announced a year after his death that his long-awaited edition of the Greek text of Rhetorius will be published in Germany and will become available in 2009. We may hope that some of his other editions may be published in the future.

Early in the year 2005, Prof. Pingree announced that he planned to retire at the end of the year. However, his health deteriorated, and late in the year he was disappointed to learn that the university planned to shut down his department after his retirement. His department in fact had consisted of only himself as faculty and a few graduate students, since the History of Mathematics is not a popular academic subject. And on 11 November 2005 he passed away. Astrologers regret the loss of an individual who was probably the foremost academic authority on early astrology, and who did a lot to make some of the early astrological works available in scholarly editions.

Some of Prof. Pingree's books on astrological subjects are:

The Thousands of Abû Ma<shar.
[fragments of Albumasar's world history]
London: The Warburg Institute, 1968.

Albumasaris de revolutionibus nativitatvm.
[Albumasar on the Revolutions of Nativities]
[an edition of the Greek text with Latin notes]
Leipzig: B. G. Teubner, 1968.

Kennedy, E. S. and David Pingree
The Astrological History of Mâshâ>allâh.
[contains facsimiles of an Arabic MS with a
lengthy citation from Mâshâ>allâh's History,
with an English translation and commentary]
Cambridge, Mass.: Harvard University Press, 1971.

Dorothei Sidonii Carmen Astrologicum.
[Dorotheus of Sidon's Astrological Poem]
[an edition of the Arabic version and the Greek
fragments, with an English translation of the Arabic]
Leipzig: B. G. Teubner, 1976.

The Yavanajâtaka of Sphudjidhvaja.
[Sanskrit text, English translation, and an elaborate commentary; as an appendix, there are brief biographical sketches in English of some of the principal Latin, Greek, Arabic, and Sanskrit writers on astrology]
Cambridge, Mass.: Harvard University Press, 1978. 2 vols.

Vettii Valentis Antiocheni Anthologiarum libri novem.
[The Nine Books of Anthology of Vettius Valens of Antioch]
[an edition of the Greek text with Latin notes and a partial edition of a relevant portion of the *Liber Hermetis*]
Leipzig: B. G. Teubner, 1986.

Raymond H. Webb MAFA, FFAA (1928-2008)

Prominent Australian astrologer. He was born at Sydney, Australia, on 13 January 1928 at 0:35 AM AEST and died 12 January 2008 at 11:40 AM AEDT at Katoomba, Australia.

Ray was a great friend and supporter of Astrology and Astrologers. His generosity of spirit has been an inspiration to me during the 28 years I have known him. As Editor and Publisher of the *Astrological Monthly Review* since taking over in 1980, his promotion and mentorship of aspiring and well established astrologers and also astrological organisations in his magazine, was outstanding. Ray was a Fellow and Professional Member of the Federation of Australian Astrologers and was presented with the Jyotish Ratna Award in 1986 for services to Astrology.

His involvement with FAA included working in executive positions and convening FAA International Conventions. He was President of his State Branch for 3 terms and a life member of The Australian Astrologers Network. He played a very important role in FAA's development and history.

Ray was also an MAFA, and he lectured internationally, in particular attracting a large following at American Federation of Astrologers, Inc. Conventions. His relaxed, easy going style and ability to present the difficult subject of astrology to the public in a clear, understandable manner, made him an instant favourite. In 1996 at the AFA Orlando Florida Convention, Ray received an award for 'Most Popular Lecturer'.

A prolific astrological writer and media personality, with many years on radio and television, Ray wrote 'Your Day by the Stars' broadcast throughout Australia and New Zealand from 1988. He was author of the 36 book series 'Pan Australian Horoscopes'.

In our lifetime we can count on one hand those we can call true friends and Ray Webb was one of those rare individuals. He died the day before his 80th birthday.

Pamela Rowe, 2ndVice President AFA

Robert Warden Cooper B.S.M.E. FAFA (1916-2008)

American astrologer and corporate executive. He was born at Philadelphia, Pennsylvania 27 Feb 1916 at 10:30 AM and died 8 Aug 2008 at Scottsdale, Arizona.

Cooper was a graduate of the Mechanical Engineering Division of Drexel University in that city. He held engineering and business management positions with Elliott Lewis-York and Willard Westinghouse in Philadelphia, and with Krauss Mechanical Contracting Co. in St. Petersburg, Fla., for 25 years.

Cooper became interested in astrology in his Eagle Scout days. He joined the American Federation of Astrologers in 1962. During his years in St. Petersburg, he assisted in the organization of the Astrological Association of Florida and served as its first President. In 1970, AFA Executive Secretary Paul R. Grell (1906-1997) elected to retire, and the Board of Directors offered the post to Mr. Cooper. He accepted and succeeded Mr. Grell as the third Executive Secretary on 11 Oct 1970.

Subsequent to that time, he planned and supervised the move of AFA Headquarters from Washington to Tempe, Ariz., effected a great improvement in AFA's financial condition, and worked constantly to increase the membership. He also initiated a number of changes in the format of AFA conventions, including the Workshops and Starlight Receptions, and particularly encouraged education and research. In 1976 he was made a Fellow of the AFA. He retired as Executive Secretary on July 29, 2007 and died at Scottsdale, Arizona, the afternoon of 8 Aug 2008.

www.ingramcontent.com/pod-product-compliance
Lightning Source LLC
LaVergne TN
LVHW090409210926
848480LV00002B/516